Analysis of Step-Stress Models

Related titles

Accelerated Testing: Statistical Models, Test Plans, and Data Analysis
(ISBN 978-0-471-69736-7)

Accelerated Life Models: Modeling and Statistical Analysis
(ISBN 978-1-58488-186-5)

Advances in Degradation Modeling: Applications to Reliability, Survival Analysis, and Finance
(ISBN 978-0-8176-4923-4)

Analysis of Step-Stress Models
Existing Results and Some Recent Developments

Debasis Kundu and Ayon Ganguly

ACADEMIC PRESS

An imprint of Elsevier

Academic Press is an imprint of Elsevier
125 London Wall, London EC2Y 5AS, United Kingdom
525 B Street, Suite 1800, San Diego, CA 92101-4495, United States
50 Hampshire Street, 5th Floor, Cambridge, MA 02139, United States
The Boulevard, Langford Lane, Kidlington, Oxford OX5 1GB, United Kingdom

Library of Congress Cataloging-in-Publication Data
A catalog record for this book is available from the Library of Congress

British Library Cataloguing-in-Publication Data
A catalogue record for this book is available from the British Library

ISBN: 978-0-12-809713-7

For information on all Academic Press publications
visit our website at https://www.elsevier.com/books-and-journals

Working together
to grow libraries in
developing countries

www.elsevier.com • www.bookaid.org

Publisher: Candice Janco
Acquisition Editor: Glyn Jones
Editorial Project Manager: Edward Payne
Production Project Manager: Debasish Ghosh
Cover Designer: Greg Harris

Typeset by SPi Global, India

To the memory of my father and to my mother
DK

To my parents
AG

Contents

Preface		**ix**
Abbreviations		**xi**
Symbols		**xiii**

1	**Introduction**	**1**
	1.1 Life testing experiments and their difficulties	1
	1.2 Accelerated life testing	1
	1.3 Censoring	3
	1.4 Different forms of data	7
	1.5 Different models	9
	1.6 Organization of the monograph	15
2	**Cumulative exposure model**	**17**
	2.1 Introduction	17
	2.2 One-parameter exponential distribution	19
	2.3 Two-parameter exponential distribution	28
	2.4 Weibull distribution	37
	2.5 Generalized exponential distribution	46
	2.6 Other continuous distributions	55
	2.7 Geometric distribution	68
	2.8 Multiple step-stress model	74
3	**Other related models**	**79**
	3.1 Introduction	79
	3.2 Tempered random variable model	79
	3.3 Tempered failure rate model	95
	3.4 Cumulative risk model	99
4	**Step-stress life tests with multiple failure modes**	**105**
	4.1 Introduction	105
	4.2 SSLT in the presence of competing risks	107
	4.3 Exponential distribution: CEM	114
	4.4 Exponential distribution: CRM	120
	4.5 Weibull distribution: TFRM	122
	4.6 SSLT in the presence of complementary risks	125

5   Miscellaneous topics   **129**
   5.1   Introduction   129
   5.2   Random stress changing time model   130
   5.3   Order restricted inference   137
   5.4   Meta-analysis approach   142
   5.5   Optimal design of SSLTs   143
   5.6   Further reading   152

Bibliography   **155**
Author index   **165**
Subject index   **169**

Preface

Accelerated life testing (ALT) is an experiment in which the experimental units are subjected to stress levels higher than the usual stress level to ensure early failure. During the past few decades, an extensive amount of work has been done related to the analysis of different ALT models in the areas of reliability and reliability engineering; see for example the books by Nelson [1] and Bagdonavicius and Nikulin [2]. Step-stress testing is a special case of ALT, which enables the experimenter to change the stress levels in a sequential manner during the experiment.

While we were searching the literature related to step-stress models, we found that during the past 15 years at least six PhD theses, several MS theses, and more than 150 research papers have been published in the reliability and reliability engineering literature. Most of the work is related to the design and analysis of the different step-stress models. These are mainly based on the parametric approach. Although an extensive amount of literature is available in this particular area, not a single book is devoted to this particular topic in detail. All the existing books devote a maximum of one or two chapters related to this topic.

The main aim of this monograph is to provide a comprehensive review of the different aspects of step-stress models and related areas. Naturally, the choice of topics and examples are based in favor of our own research interests. We have tried to include almost all the references related to this area which are currently available and our main source is the Google search engine. We are sure that the list of references is far from complete, but this is not intentional.

We have kept the mathematical level quite modest throughout the book. Graduate level statistics courses should be sufficient preparation to understand the mathematics in all the chapters. We have avoided proofs in most of the cases but we have provided the relevant references. This monograph has five chapters. After a brief introduction to the topic in Chapter 1, we have discussed different models and their analyses in Chapters 2–4. In Chapter 5, we have briefly discussed several related topics and provided an extensive list of references for further reading. In each chapter we have indicated several open problems for future research.

Every book is written with a specific audience in mind. This book is not a textbook per se. It has been written mainly for graduate students specializing in mathematics, statistics, or industrial engineering and young researchers who are planning to work in the area of reliability. This book will provide an easy reference and it will be helpful for a young researcher to find a research topic in this area. We hope this book will motivate young researchers to pursue their research in this particular area. We will consider our efforts to be worthy if the target audience finds this volume useful.

Debasis Kundu
Kanpur, India

Ayon Ganguly
Guwahati, India

Abbreviations

ALT	accelerated life testing
BE	Bayes estimator/estimate
CDF	cumulative distribution function
CEM	cumulative exposure model
CMGF	conditional moment generating function
CRI	credible interval
CRM	cumulative risk model
FRF	failure rate function
GHCS-I	generalized Type-I hybrid censoring scheme
GHSC-II	generalized Type-II hybrid censoring scheme
HCS-I	Type-I hybrid censoring scheme
HCS-II	Type-II hybrid censoring scheme
HPD	highest posterior density
i.i.d.	identically and independently distributed
LL	lower limit
MGF	moment generating function
MLE	maximum likelihood estimator/estimate
PCS-I	progressive Type-I censoring scheme
PCS-II	progressive Type-II censoring scheme
PDF	probability density function
PHCS	progressive hybrid censoring scheme
PMF	probability mass function
SSLT	step-stress life test(ing)
TFRM	tampered failure rate model
TR(A)	trace of a square matrix A
TRVM	tampered random variable model
UL	upper limit

Symbols

z_p	the pth upper percentile point of the standard normal distribution
T	random variable denoting the lifetime
$T_{i:n}$	ith order statistic with sample size n
s_i	ith stress level
τ_i	time at which the stress is changed from s_i to s_{i+1}
η	Type-I censoring time
$f_X(x; \boldsymbol{\theta})$	probability density function of the random variable X at the point x having parameter $\boldsymbol{\theta}$
$F_X(x; \boldsymbol{\theta})$	distribution function of the random variable X at the point x having parameter $\boldsymbol{\theta}$
$\overline{F}_X(x; \boldsymbol{\theta})$	$1 - F_X(x; \boldsymbol{\theta})$.
Beta(a, b)	beta distribution having PDF $\frac{\Gamma(a+b)}{\Gamma(a)\Gamma(b)} x^{a-1}(1-x)^{b-1}$ for $0 < x < 1$
Bin(n, p)	binomial distribution with parameters n and p
Exp(θ)	exponential distribution with mean θ
Exp(μ, θ)	exponential distribution having PDF $(1/\theta) e^{-(x-\mu)/\theta}$ for $x > \mu$
Wei(α, λ)	Weibull distribution having PDF $\alpha\lambda x^{\alpha-1}e^{-\lambda x^\alpha}$ for $x > 0$
Gamma(α, λ)	gamma distribution having PDF $\left(\lambda^\alpha / \Gamma(\alpha)\right) x^{\alpha-1}e^{-\lambda x}$ for $x > 0$
IGamma(α, λ)	inverse gamma distribution having PDF $\frac{\lambda^\alpha}{\Gamma(\alpha)} e^{-\frac{\lambda}{x}} \left(\frac{1}{x}\right)^{\alpha+1}$ if $x > 0$
U(a, b)	continuous uniform random variable over the interval (a, b)
GE(θ)	geometric distribution with the PMF $P(X = x; \theta) = \theta(1-\theta)^{x-1}$ for $x = 1, 2, \ldots$
$\mathbb{1}_A(\cdot)$	indicator function of the set A
$\Gamma(a)$	complete gamma function; $\int_0^\infty t^{a-1}e^{-t}dt$
$\Gamma(a, z)$	incomplete gamma function; $\frac{1}{\Gamma(a)} \int_z^\infty t^{a-1}e^{-t}dt$
$\Phi(x)$	CDF of standard normal distribution at the point x
$\phi(x)$	PDF of standard normal distribution at the point x
$\langle x \rangle$	$\max\{x, 0\}$

Introduction

1

1.1 Life testing experiments and their difficulties

Life testing experiments have gained a significant amount of popularity in recent times. The main aim of any life testing experiment is to measure one or more reliability characteristics of the experimental units under consideration. In a very classical form of a life testing experiment, a certain number of identical items are placed on the test under normal operating conditions and the "time to failure" of all the items is recorded. The definition of the "time to failure" depends on the items considered. For example, "time to failure" may be the time after which a minimum satisfactory performance is not achieved for a piece of electronic equipment, or it may be the number of revolutions before a malfunctioning of a ball bearing. For testing the lifetime of an electric bulb, "time to failure" is the number of hours it works before it is fused. The failure may occur due to any one or a combination of more than one of the following reasons: (a) careless planning, (b) substandard raw materials, (c) wear-out or fatigue caused by the aging of the item, etc. As the failure can occur at any time, it is assumed that the "time to failure" is a random variable having a specific cumulative distribution function (CDF).

Due to substantial improvement of the science and technology, most of the industrial products available today are extremely reliable with large mean times to failure under their normal operating conditions. Consequently, it may not be possible to obtain adequate information about the lifetime distributions and the associated parameters within an affordable time using conventional life testing experiments. Moreover, most of the life testing experiments are destructive in nature, i.e., items put on test cannot be used for future purposes. Due to these problems, the reliability experimenter may resort to accelerated life testing (ALT) and/or different censoring techniques, as will be described next.

1.2 Accelerated life testing

In an ALT experiment, the experimental units are subjected to higher stress levels than the normal operating conditions. It affects the lifetime of the items under consideration negatively, hence the items fail quickly than under the normal conditions. The factors that affect the lifetime of an item are called stress factors. For example, voltage,

Analysis of Step-Stress Models. http://dx.doi.org/10.1016/B978-0-12-809713-7.00001-6

temperature, and humidity could be stress factors for electronic equipment. Electronic products such as toasters, washers, electronic chips, etc. are expected to last over a period of time much longer than what laboratory testing would allow. Therefore, using the ALT experiment one can obtain valuable information about the product reliability within the experimental time limits. The ALT experiment may be performed either at a constant high stress level or different stress levels. The data obtained from an ALT experiment are used to draw conclusions about the parameters of the lifetime distribution under normal operating conditions.

A special case of the ALT experiment is the step-stress life test (SSLT), which enables the experimenter to change the level of the stress factors in a sequential manner during the experiment. Let $s_1, \ldots, s_k$ be k predetermined stress levels and $\tau_1 < \cdots < \tau_{k-1}$ be $(k-1)$ prespecified time points. In a very basic form of SSLT, n units are put on the test at an initial stress level s_1. At the time point τ_1, the stress level is changed to s_2 from s_1. Similarly at the time point τ_2, the stress level is changed from s_2 to s_3 and so on. Finally at the time point τ_{k-1}, the stress level is changed to s_k from s_{k-1}. Therefore, if $s(t)$ denotes the stress level at the time point t, then

$$
s(t) = \begin{cases} s_1 & \text{if} \quad \tau_0 \leq t < \tau_1, \\ s_2 & \text{if} \quad \tau_1 \leq t < \tau_2, \\ \vdots & \vdots \qquad \vdots \\ s_k & \text{if} \quad \tau_{k-1} \leq t < \tau_k, \end{cases}
$$

where $\tau_0 = 0$ and $\tau_k = \infty$. The experiment stops when all the items put on test fail. This is also known as the fixed stress changing time SSLT.

The failure times are recorded in chronological order. If we assume that the number of failures before the time τ_i, for $i = 1, \ldots, k - 1$, is n_i, then a typical complete data set looks like

$$
t_{1:n} < \cdots < t_{n_1:n} < \tau_1 < t_{n_1+1:n} < \cdots < t_{n_2:n} < \tau_2 < \cdots < \tau_{k-1} < t_{n_{k-1}+1:n} < \cdots < t_{n:n}.
$$

A simple SSLT is a special case of a SSLT when it involves only two stress levels s_1 and s_2, and the stress change takes place at a prefixed time point τ_1. A simple step stress model has been discussed quite extensively in the literature under various model assumptions for different lifetime distributions. We will be discussing the analysis of different simple step stress models and related issues in the subsequent chapters.

Alternatively, instead of changing the stress levels at prefixed time points, the stress levels can be changed at random time points also. For example, n items are put on life testing experiments at the initial stress level s_1. Let $r_1, r_2, \ldots, r_k$ be prefixed positive integers such that $1 < r_1 < \cdots < r_{k-1} < n$. As before, the failure times are recorded in a chronological manner. At the time of the r_1th failure, the stress level is changed from s_1 to s_2. Similarly, at the time of the r_2th failure, the stress level is changed from s_2 to s_3, and so on. Finally, at the time of the r_{k-1}th failure, the stress level is changed from s_{k-1} to s_k. This is known as the random stress changing time SSLT experiment. In this case a typical complete data set will be as follows:

$$t_{1:n} < \cdots < t_{r_1:n} < t_{r_1+1:n} < \cdots < t_{r_2:n} < \cdots < t_{r_{k-1}+1:n} < \cdots < t_{n:n},$$

where it is known that the stress levels have been changed at the random time points $t_{r_1:n} < t_{r_2:n} < \cdots < t_{r_{k-1}:n}$.

1.3 Censoring

Censoring is inevitable in most of the life testing experiments. Censoring basically means terminating the experiment in a well-planned manner before the failure of all the items put into a test. Censoring can be done with respect to a prespecified time or a prespecified number of failures or a combination of both. Depending upon the censoring criteria there are different types of censoring schemes available in the literature. Consider the following experiment. Let n be a positive integer, and a total of n items are put into a life testing experiment. Let $t_{1:n} < t_{2:n} < \cdots < t_{n:n}$ be the ordered failure times of the items. Throughout it is assumed that the failed items are not replaced. Now we will discuss different popular censoring schemes which are used in practice.

1.3.1 Basic censoring schemes

Type-I and Type-II censoring schemes are the two most common and popular censoring schemes. They are described as follows.

Type-I censoring scheme

Let η be a prefixed time. In a Type-I censoring scheme the experiment is stopped at the time point η. Hence under this censoring scheme, the experimental time cannot exceed η, and the data set is one of the following forms.

(a) $t_{1:n} < t_{2:n} < \cdots < t_{d:n} < \eta$,
(b) $t_{1:n} < t_{2:n} < \cdots < t_{n:n} < \eta$,
(c) there is no failure before the time η,

where $d \in \{1, \ldots, n\}$ is the number of failures before the time η. Therefore, in this case although the experimental time is fixed, the number of failures is a random variable taking values $0, 1, \ldots, n$. Clearly, prefixed experimental duration is the main advantage of a Type-I censoring scheme, although a wrongly chosen η may result in very few or even no failures before the experiment stops. If there are few failures, the inference based on a small sample may not be efficient. Although statistical inference may be possible in case of no failure, the results may not be very informative; see Meeker and Escobar [3] and Nelson [1]. This is a major drawback of a Type-I censoring scheme.

Interested readers are referred to Lawless [4], Miller [5], and Bain and Englehardt [6] in this respect.

Type-II censoring scheme

Let r ($\leq n$) be a prefixed positive integer. In a Type-II censoring scheme the experiment is terminated as soon as the rth failure takes place. Under a Type-II censoring scheme the data set looks like

(a) $t_{1:n} < t_{2:n} < \cdots < t_{r:n}$.

In contrast to the Type-I censoring scheme, in this case the number of failures is prefixed, but the experimental time is a random variable. Clearly, the prefixed number of failures is the main advantage, whereas the unbounded experimental duration is the main disadvantage of a Type-II censoring scheme. Interested readers are referred to Lawless [4], Miller [5], and Bain and Englehardt [6] for more detailed discussions on Type-II censoring scheme.

1.3.2 Hybrid and generalized hybrid censoring schemes

A hybrid censoring scheme (HCS) is a mixture of Type-I and Type-II censoring schemes. Now we will describe briefly the hybrid Type-I censoring scheme (HCS-I), hybrid Type-II censoring scheme (HCS-II), generalized hybrid Type-I censoring scheme (GHCS-I), and generalized hybrid Type-II censoring scheme (GHCS-II).

Hybrid Type-I censoring scheme

Epstein [7] first introduced HCS-I which can be described as follows. Let r ($\leq n$) be a prefixed positive integer, and η be a predetermined time. The test is terminated when the rth item fails or time η is reached, whichever is earlier, i.e., the termination time of the experiment is $\eta_* = \min\{t_{r:n}, \eta\}$. For a HCS-I, the available data will be of the form

(a) $t_{1:n} < t_{2:n} < \cdots < t_{r:n}$ if $\eta_* = t_{r:n}$,
(b) $t_{1:n} < t_{2:n} < \cdots < t_{d:n}$ if $\eta_* = \eta$,
(c) there is no failure before the time η,

where $d \in \{1, 2, \ldots, r - 1\}$ is the number of failures before the time η. Note that the maximum duration of the experiment under this censoring scheme is η and this is the main advantage of this censoring scheme. Like the Type-I censoring scheme, the experiment can be terminated with very few or no failures before the time η, and this is a serious drawback of the HCS-I.

Hybrid Type-II censoring scheme

To overcome the disadvantage of the HCS-I by ensuring a minimum number of failures, Childs et al. [8] introduced HCS-II, and it can be described as follows. Let r ($\leq n$) be a prefixed positive integer, and η be a predetermined time as defined before in HCS-I. The test is terminated when the rth failure occurs or the time η is reached, whichever is later, i.e., the termination time of the experiment is $\eta^* = \max\{t_{r:n}, \eta\}$. For a HCS-II, the available data will be of the form

(a) $t_{1:n} < t_{2:n} < \cdots < t_{r:n}$ if $\eta^* = t_{r:n}$,
(b) $t_{1:n} < t_{2:n} < \cdots < t_{d:n}$ if $\eta^* = \eta$,

where $d \in \{r+1, r+2, \ldots, n\}$ is the number of failures before the time η. Note that in this case the number of failures is restricted to the set $\{r, r+1, \ldots, n\}$. Therefore, for a HCS-II, it is ensured that the experimenter is going to observe at least r failures. However, for a HCS-II there is no upper bound on the time duration of the experiment, which is the main disadvantage of this censoring scheme.

Generalized hybrid Type-I censoring scheme

Chandrasekar et al. [9] introduced GHCS-I and GHCS-II mainly to overcome the drawbacks of HCS-I and HCS-II. The GHCS-I can be described as follows. Let r and k be two prefixed positive integers satisfying $k < r \le n$ and η be a predetermined time. If the kth failure occurs before the time η, the experiment is terminated at $\eta_* = \min\{t_{r:n}, \eta\}$. If the kth failure occurs after time point η, the experiment is terminated at the time $t_{k:n}$. Hence, GHCS-I modifies HCS-I by allowing the experiment to run beyond the time η. For a GHCS-I, the available data will be of the following form

(a) $t_{1:n} < t_{2:n} < \cdots < t_{k:n} < \cdots < t_{r:n}$ if $t_{r:n} \le \eta$,
(b) $t_{1:n} < t_{2:n} < \cdots < t_{k:n} < \cdots < t_{d:n}$ if $t_{k:n} < \eta < t_{r:n}$,
(c) $t_{1:n} < \cdots < t_{k:n}$ if $t_{k:n} > \eta$.

Here $d \in \{k, k+1, \ldots, r-1\}$ is the number of failures before the time η. Therefore, for a GHCS-I, the number of failures is restricted between k and r. Under this censoring scheme the experimenter would like to observe r failures, but is willing to accept a bare minimum of k failures. Although theoretically there is no upper bound on the time duration of the experiment, for a proper choice of k, the expected time duration of the experiment can be controlled; see Chandrasekar et al. [9] for details.

Generalized hybrid Type-II censoring scheme

The GHCS-II was also proposed by Chandrasekar et al. [9], and it can be described as follows. Let r $(\le n)$ be a prefixed positive integer and $\eta_1 > 0$, $\eta_2 > 0$ be two prespecified times such that $\eta_1 < \eta_2$. If the rth failure occurs before time point η_1, the experiment is terminated at the time point η_1. If the rth failure occurs between η_1 and η_2, the experiment stops at the time $t_{r:n}$. Otherwise, the experiment is terminated at the time η_2. Thus GHCS-II modifies the HCS-II by reducing the experimental time to η_2 from above. For a GHCS-II, the available data will be of the following form

(a) $t_{1:n} < \cdots < t_{r:n} < \cdots < t_{d_1:n}$ if $t_{r:n} \le t_{d_1:n} < \eta_1$,
(b) $t_{1:n} < t_{2:n} < \cdots < t_{r:n}$ if $\eta_1 < t_{r:n} < \eta_2$,
(c) $t_{1:n} < \cdots < t_{d_2:n} < \eta_2$ if $t_{r:n} > \eta_2$.

Here $d_1 \in \{r, r+1, \ldots, n\}$ is the number of failures before the time η_1 when the rth failure occurs before η_1 and $d_2 \in \{0, 1, \ldots, r-1\}$ is the number of failures before the time η_2 when the rth failure occurs after η_2. Therefore, for a GHCS-II, the experimental time is restricted between η_1 and η_2. In this censoring scheme, the experimenter would like to observe r failures, but is willing to continue the experiment till the time η_1, if the rth failure occurs before η_1. On the other hand, if the rth failure does not take place before η_2, the experimenter does not want to continue the experiment beyond η_2. Interested readers are referred to the review article by Balakrishnan and Kundu [10] in this regard.

1.3.3 Progressive censoring scheme

None of the censoring schemes that have been discussed allows removal of the experimental units during the experiment. The progressive censoring scheme (PCS) allows the experimenter to remove experimental units during the experiment. The removed items may be used for other purposes. Further, the PCS has some other advantages. For example, it provides more information about the tail behavior of the lifetime distribution of the experimental units. Extensive work has been done on PCS in the last two decades. The recent book by Balakrishnan and Cramer [11] provided detailed descriptions of all the developments related to PCS. Different PCSs are available in practice, and we will briefly describe them in the following paragraphs.

Progressive Type-I censoring scheme

Let m ($\leq n$) be a prespecified positive integer. Let $\eta_1 < \eta_2 < \cdots < \eta_m$ be m predetermined time points, and $R_1, \ldots, R_{m-1}$ be prefixed nonnegative integers. Let N_1 be the number of failures before the time η_1. R_1 items are chosen at random from the remaining $(n - N_1)$ surviving units and removed from the test at the time point η_1. The experiment continues, and suppose N_2 is the number of failures between the time points η_1 and η_2. Out of the $(n - N_1 - R_1 - N_2)$ surviving units, R_2 items are chosen at random and removed at the time point η_2, and so on. Finally, at the time point η_m, all the remaining items, say R_m, are censored and the experiment is terminated. This censoring scheme is known as the progressive Type-I censoring scheme (PCS-I). Note that a PCS-I is feasible if the number of units still on the test at each censoring time is larger than the number of items planned to be censored at that time point, and the feasibility of such a censoring scheme is always assumed; see for example Balakrishnan [12]. Clearly, the termination time of the experiment is fixed at η_m, and we have the relation $\sum_{j=1}^{m} (N_j + R_j) = n$. The experimental time of the PCS-I cannot exceed η_m, but the number of failures in a PCS-I is a random variable. The available data from a PCS-I will be of the following form:

(a) $t_{1:n} < \cdots < t_{n_1:n} < \eta_1 < \cdots < \eta_{m-1} < t_{n_1 + \cdots + n_{m-1} + 1} < \cdots < t_{n_1 + \cdots + n_{m-1} + n_m} < \eta_m.$

Progressive Type-II censoring scheme

Let m ($\leq n$) be a prespecified positive integer. Let $R_1, R_2, \ldots, R_m$ be m predetermined nonnegative integers satisfying $m + \sum_{j=1}^{m} R_j = n$. At the point of the first failure $t_{1:n}$, R_1 items are selected at random from the remaining $(n - 1)$ surviving units and they are removed. Similarly, at the time of the second failure $t_{2:n}$, R_2 items are selected at random from the remaining $(n - 2 - R_1)$ surviving units and they are removed, and so on. Finally, at the time of the mth failure $t_{m:n}$, the remaining R_m units are removed from the experiment and the experiment stops. This particular censoring scheme is known as the progressive Type-II censoring scheme (PCS-II). The number of failures of a PCS-II is fixed, whereas the experimental time is not bounded. The available data from a PCS-II will be of the form:

(a) $t_{1:n} < t_{2:n} < \cdots < t_{m:n}$.

Progressive hybrid censoring scheme

Let m ($\leq n$) be a prefixed positive integer, $R_1, R_2, \ldots, R_m$ be m predetermined nonnegative integers such that $m + \sum_{j=1}^{m} R_j = n$. Let η be a predetermined time. At the time of the first failure $t_{1:n}$, similarly as before, R_1 items are chosen at random from the remaining $(n - 1)$ surviving units and removed from the experiment. Similarly, at the time of the second failure $t_{2:n}$, out of the remaining $(n - 2 - R_1)$ surviving units R_2 items are chosen at random and removed, and so on. If the mth failure occurs before time point η, the remaining R_m units are removed from the experiment and the experiment stops at the time point $t_{m:n}$. On the other hand if there are fewer failures than m before the time point η, the experiment is terminated at the time point η by removing all the remaining items from the test. In this case, the number of censored units at the time η is $R_N^* = n - N - \sum_{j=1}^{N} R_j$, where $0 \leq N < m$ is the number of failures before the time point η. Hence, under the progressive hybrid censoring scheme (PHCS) the observed data will be of the following form:

(a) $t_{1:n} < \cdots < t_{m:n}$ if $t_{m:n} < \eta$,
(b) $t_{1:n} < \cdots < t_{d:n}$ if $t_{d:n} < \tau < t_{d+1:n} \leq t_{m:n}$.

Here $d \in \{1, 2, \ldots, m - 1\}$ is the number of failure befor the time η.

For a detailed description of the PHCS, the readers are referred to Kundu and Joarder [13] or Childs et al. [14].

1.4 Different forms of data

In Section 1.2 we have discussed the simple SSLT experiment. In this section we provide various forms of data obtained from a simple SSLT experiment under different censoring schemes. It is assumed that a total of n units is placed on a simple SSLT experiment. The stress level is changed from s_1 to s_2 at a prefixed time point τ_1, and $\eta > \tau_1$ is another prefixed time. The positive integer r ($\leq n$) is also prefixed. The role of r and η will be clear soon. Let the ordered lifetime of the items be denoted by

$t_{1:n} < \cdots < t_{n:n}$. Further, let n_1 and n_2 denote the number of failures before the time point τ_1 and between the time points τ_1 and η, respectively.

Type-I censoring scheme

The test is terminated at the time point η. For a Type-I censoring scheme the available data is one of the following forms:

(a) $\tau_1 < t_{1:n} < \cdots < t_{n_2:n} < \eta; \quad 0 < n_2 \le n.$
(b) $t_{1:n} < \cdots < t_{n_1:n} < \tau_1 < t_{n_1+1} < \cdots < t_{n_1+n_2:n} < \eta; \; n_1 > 0, n_2 > 0, n_1 + n_2 \le n.$
(c) $t_{1:n} < \cdots < t_{n_1:n} < \tau_1 < \eta.$

In case (a) there is no failure before τ_1 ($n_1 = 0$) and the experiment stops at the time point η if $n_2 < n$, otherwise it stops at $t_{n:n}$. In case (b) the experiment stops at η if $n_1 < n$, otherwise it stops at $t_{n:n}$. In case (c) there is no failure between τ_1 and η ($n_2 = 0$), and the experiment stops at η. It may be possible that no failure takes place before η, and hence no data is available from the experiment.

Type-II censoring scheme

For a Type-II censoring scheme the test is terminated at the time of the rth failure. Therefore, the test is terminated at a random time $t_{r:n}$. In this case the available data is one of the following forms:

(a) $\tau_1 < t_{1:n} < \cdots < t_{r:n}.$
(b) $t_{1:n} < \cdots < t_{n_1:n} < \tau_1 < t_{n_1+1} < \cdots < t_{r:n}; \; 0 < n_1 < r.$
(c) $t_{1:n} < \cdots < t_{r:n} < \tau_1.$

In case (a) there is no failure before τ_1 ($n_1 = 0$) and in case (c) the experiment stops before τ_1 ($n_1 = r$).

Hybrid Type-I censoring scheme

In this case, the experiment terminates at a random time $\tau_* = \min\{t_{r:n}, \eta\}$. For HCS-I, the available data is one of the following forms:

(a) $\tau_1 < t_{1:n} < \cdots < t_{r:n} < \eta.$
(b) $t_{1:n} < \cdots < t_{n_1:n} < \tau_1 < t_{n_1+1} < \cdots < t_{r:n} < \eta; \;$ if $\; 0 < n_1 < r.$
(c) $t_{1:n} < \cdots < t_{r:n} < \tau_1.$
(d) $\tau_1 < t_{1:n} < \cdots < t_{n_2:n} < \eta; \;$ if $\; t_{r:n} > \eta.$
(e) $t_{1:n} < \cdots < t_{n_1:n} < \tau_1 < t_{n_1+1:n} < \cdots < t_{n_1+n_2:n} < \eta; \;$ if $\; t_{r:n} > \eta, n_1 < r.$
(f) $t_{1:n} < \cdots < t_{n_1:n} < \tau_1 < \eta; \;$ if $\; t_{r:n} > \eta.$

In case (a) there is no failure before τ_1 ($n_1 = 0$) and the experiment stops at $t_{r:n}$ before the time point η. In case (b) the rth failure takes place between the time points τ_1 and η and the experiment stops at $t_{r:n}$. In case (c), the experiment stops at $t_{r:n}$ before the time point τ_1. In cases (d), (e), and (f), the rth failure takes place beyond η, and in all the cases the experiment stops at η.

Hybrid Type-II censoring scheme

In this censoring scheme, the experiment terminates at a random time $\tau^* = \max\{t_{r:n}, \eta\}$. For an HCS-II, the available data from a simple SSLT is one of the following forms:

(a) $\tau_1 < t_{1:n} < \cdots < t_{r:n};$ if $t_{r:n} < \eta.$

(b) $t_{1:n} < \cdots < t_{n_1:n} < \tau_1 < t_{n_1+1} < \cdots < t_{r:n};$ if $t_{r:n} > \eta, 0 < n_1 < r.$

(c) $t_{1:n} < \cdots < t_{n_1:n} < \tau_1 < \eta;$ if $n_1 \geq r.$

(d) $\tau_1 < t_{1:n} < \cdots < t_{n_2:n} < \eta;$ if $t_{r:n} < \eta.$

(e) $t_{1:n} < \cdots < t_{n_1:n} < \tau_1 < t_{n_1+1:n} < \cdots < t_{r:n} < \cdots < t_{n_1+n_2:n} < \eta;$ if $t_{r:n} < \eta.$

In case (a) there is no failure before τ_1 ($n_1 = 0$) and the experiment stops at $t_{r:n}$, beyond the time point η. In case (b) the rth failure takes place beyond the time point η and the experiment stops at $t_{r:n}$. In case (c), the experiment stops at η beyond τ_1, and no failure takes place at the second stress level. In cases (d) and (e) the rth failure takes place before η, and the experiment stops at η.

Progressive Type-II censoring scheme

Let $R_1, \ldots, R_m$ be m prefixed nonnegative integers such that

$$m + \sum_{i=1}^{m} R_i = n.$$

The experiment stops at $t_{m:n}$, when the mth failure takes place. As has been discussed before, in a PCS-II, $R_1, \ldots, R_m$ surviving units are removed at the time point $t_{1:n} < \cdots < t_{m:n}$, respectively, from the experiment. In this case the data is one of the following forms:

(a) $\tau_1 < t_{1:n} < \cdots < t_{m:n}.$

(b) $t_{1:n} < \cdots < t_{n_1:n} < \tau_1 < t_{n_1+1:n} < \cdots < t_{m:n},$ if $n_1 < m.$

(c) $t_{1:n} < \cdots < t_{m:n} < \tau_1.$

In case (a) there is no failure before τ_1, and all the failure takes place at the stress level s_2, and in case (c) all the failure takes place before τ_1.

1.5   Different models

Let us assume that the CDF of the lifetime at the stress level s_i is $F_i(\cdot)$. To analyze data under an SSLT, one needs a model that relates the CDFs of the lifetime under different stress levels to the CDF of the lifetime of the product under the normal operating condition. Several models have been proposed in the literature to describe this relationship.

1.5.1 Cumulative exposure model

Among the different models the most popular and commonly used one is known as the cumulative exposure model (CEM), first proposed by Sediakin [15] and later extensively studied by Nelson [16]. The CEM assumes that the residual life of the experimental units depends only on the cumulative exposure the units have experienced, with no memory of how this exposure was accumulated. Moreover, at a fixed stress level units fail according to the CDF of that stress level, but the overall CDF maintains the continuity. To develop the necessary conditions required for a CEM, let us consider a general k-step SSLT, where the stress levels are changed at the prefixed time point $\tau_1 < \cdots < \tau_{k-1}$. Let $F(\cdot)$ denote the CDF of the lifetime of an experimental unit from the previous k-step SSLT experiment.

Clearly, under the assumption of the CEM, the units fail according to the CDF of the stress level s_1 till the time point τ_1, and hence

$$F(t) = F_1(t) \quad \text{for} \quad 0 \leq t \leq \tau_1.$$

The effect of change of the stress level from s_1 to s_2 at the time point τ_1 is equivalent to change the CDF of the stress level s_2 from $F_2(t)$ to $F_2(t - h_1)$, i.e.,

$$F(t) = F_2(t - h_1) \quad \text{for} \quad \tau_1 \leq t \leq \tau_2,$$

while the shifting parameter h_1 can be obtained as the solution of the equation

$$F_2(\tau_1 - h_1) = F_1(\tau_1),$$

because of the continuity of the function $F(t)$ at τ_1. Proceeding in this way, with $\tau_0 = 0$ and $\tau_k = \infty$, finally one will have

$$F(t) = F_i(t - h_i) \quad \text{for} \quad \tau_{i-1} \leq t \leq \tau_i, \quad i = 1, \ldots, k, \tag{1.1}$$

where $h_0 = 0$ and h_i for $i = 1, 2, \ldots, k - 1$, is the solution of the equation

$$F_{i+1}(\tau_i - h_i) = F_i(\tau_i - h_{i-1}).$$

Let us consider a specific example of the CEM. Here it is assumed that $k = 3$, and the lifetime of the experimental units follow exponential distribution with the scale parameter (mean) θ_i at the stress level s_i, for $i = 1, 2, 3$. Hence the CDF of the lifetime at the stress level s_i is given by

$$F_i(t) = \begin{cases} 0 & \text{if} \quad t < 0 \\ 1 - e^{-\frac{t}{\theta_i}} & \text{if} \quad t \geq 0. \end{cases}$$

The CDF of the lifetime under the assumption of the CEM can be obtained as follows. Till the time point τ_1, it is the same as the CDF of the lifetime at the stress level s_1. For $t \in (\tau_1, \tau_2)$, $F(t) = F_2(t - h_1)$, where h_1 is the solution of the equation

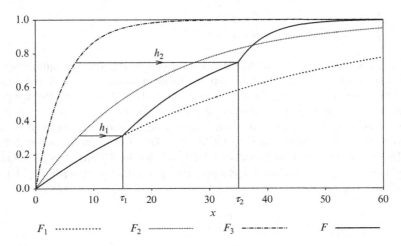

Fig. 1.1 Pictorial presentation of CDF under CEM.

$$1 - e^{-\frac{\tau_1}{\theta_1}} = 1 - e^{-\frac{\tau_1 - h_1}{\theta_2}}.$$

Hence $h_1 = (1 - \theta_2/\theta_1)\tau_1$. Similarly, for $t \in (\tau_2, \infty)$, the CDF of the lifetime under the SSLT is given by $F(t) = F_3(t - h_2)$, where h_2 is the solution of the equation

$$1 - e^{-\frac{\tau_2 - h_2}{\theta_3}} = 1 - e^{-\frac{\tau_2 - h_1}{\theta_2}}.$$

Hence we obtain $h_2 = (1 - \theta_3/\theta_2)\tau_2 + (1/\theta_2 - 1/\theta_1)\theta_3\tau_1$. Therefore, we have

$$F(t) = \begin{cases} 0 & \text{if } t < 0 \\ 1 - e^{-\frac{t}{\theta_1}} & \text{if } 0 \le t < \tau_1 \\ 1 - e^{-\frac{t-\tau_1}{\theta_2} - \frac{\tau_1}{\theta_1}} & \text{if } \tau_1 \le t < \tau_2 \\ 1 - e^{-\frac{t-\tau_2}{\theta_3} - \frac{\tau_2-\tau_1}{\theta_2} - \frac{\tau_1}{\theta_1}} & \text{if } t \ge \tau_2. \end{cases} \tag{1.2}$$

The process of obtaining the CDF under the assumption of CEM is depicted in Fig. 1.1, where we have taken $\theta_1 = 40$, $\theta_2 = 20$, $\theta_3 = 5$, $\tau_1 = 15$, and $\tau_2 = 35$.

1.5.2 Tampered random variable model

The tampered random variable model (TRVM) was first proposed by Goel [17]; see also Goel [18] and DeGroot and Goel [19]. Consider a simple SSLT. It is assumed that the effect of the change of the stress level from s_1 to s_2 at the time τ_1 is equivalent to multiply the remaining life of the experimental unit by an unknown positive constant, say β, which depends on both the stress levels. Mathematically, if T denotes the lifetime under the stress level s_1, then the lifetime $\widetilde{T}$ for a simple SSLT model for a TRVM assumption is given by

$$\widetilde{T} = \begin{cases} T & \text{if } T \le \tau_1 \\ \tau_1 + \beta^{-1}(T - \tau_1) & \text{if } T > \tau_1. \end{cases} \tag{1.3}$$

Because the switching to the higher stress level can be regarded as tampering with the ordinary lifetime, $\widetilde{T}$ is called a tampered random variable. Here β is known as the acceleration or tampering factor. Usually $\beta > 1$. β^{-1} is called the tampering coefficient and τ_1 is called the tampering point. If the observed value of $\widetilde{T}$ is less than the tampering point, it is called untampered observation; otherwise it is called tampered observation. Note that for a CEM, the CDF of the lifetime under different stress levels may be completely unrelated. However, for the TRVM, they are related to each other by Eq. (1.3). Suppose T has a probability density function (PDF) $f_T(\cdot)$, then the PDF of $\widetilde{T}$ is given by

$$f_{\widetilde{T}}(t) = \begin{cases} f_T(t) & \text{if } t \le \tau_1 \\ \beta f_T(\tau_1 + \beta(t - \tau_1)) & \text{if } t > \tau_1. \end{cases} \tag{1.4}$$

Let us consider a specific example. Suppose T is a Weibull random variable with the shape and the scale parameters α and θ, respectively. The PDF of T is given by

$$f_T(t) = \begin{cases} \frac{\alpha}{\theta} t^{\alpha-1} e^{-t^\alpha/\theta} & \text{if } 0 < t < \infty \\ 0 & \text{otherwise.} \end{cases}$$

Hence, using Eq. (1.4), the PDF of $\widetilde{T}$ becomes

$$f_{\widetilde{T}}(t) = \begin{cases} \frac{\alpha}{\theta} t^{\alpha-1} e^{-t^\alpha/\theta} & \text{if } 0 < t \le \tau_1 \\ \frac{\beta\alpha}{\theta} (\tau_1 + \beta(t - \tau_1))^{\alpha-1} e^{-(\tau_1+\beta(t-\tau_1))^\alpha/\theta} & \text{if } t > \tau_1 \\ 0 & \text{otherwise;} \end{cases}$$

The process of obtaining the CDF under the assumption of a TRVM is depicted in Fig. 1.2. Here we have taken $\alpha = 1.5$, $\theta = 4$, $\beta = 3$, and $\tau = 3$. It can be shown

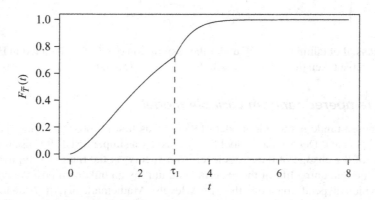

Fig. 1.2 Pictorial presentation of CDF under TRVM.

that the TRVM becomes equivalent to CEM if the lifetime distributions under the two stresses belong to the same scale family; see Rao [20].

1.5.3 Tampered failure rate model

The tampered failure rate model (TFRM) was proposed by Bhattacharyya and Soejoeti [21] for a simple SSLT. Borrowing the idea of Cox's hazard rate model (see Cox [22]), Bhattacharyya and Soejoeti [21] assumed that the effect of switching the stress level from s_1 to s_2 at the time τ_1 is to multiply the failure rate function (FRF) at the stress level s_1 by an unknown constant β, i.e.,

$$\tilde{\lambda}(t) = \begin{cases} \lambda(t) & \text{if } t \le \tau_1 \\ \beta\lambda(t) & \text{if } t > \tau_1 \end{cases} \tag{1.5}$$

where $\lambda(\cdot)$ and $\tilde{\lambda}(\cdot)$ are the failure rates at the stress level s_1 and that under a simple SSLT, respectively. The factor β depends on both the stress levels s_1 and s_2 and possibly on τ_1. However, the authors assumed that β does not depend on t. Like the TRVM, the CDFs at the stress levels s_1 and s_2 under the TFRM assumption are related to each other by Eq. (1.5). Suppose the CDF at the stress level s_1 is $F(\cdot)$; then the CDF of the lifetime of an experimental unit from this step stress pattern is given by

$$\tilde{F}(t) = \begin{cases} F(t) & \text{if } t \le \tau_1 \\ 1 - (1 - F(\tau_1))^{1-\beta} (1 - F(t))^{\beta} & \text{if } t > \tau_1. \end{cases} \tag{1.6}$$

It can be easily seen that the TRVM and TFRM will coincide if $\beta = 1$ or the lifetime distribution has a lack of memory property; see Rao [20]. Madi [23] proposed the TFRM for multiple step SSLT generalizing the concept of Bhattacharyya and Soejoeti [21]. Madi [23] assumed that the failure rate, $\tilde{\lambda}(t)$, under a multiple SSLT is given by

$$\tilde{\lambda}(t) = \left(\prod_{j=0}^{i-1} \beta_j\right) \lambda(t) \quad \text{if } \tau_{i-1} \le t < \tau_i \quad i = 1, \dots, k, \tag{1.7}$$

where $\tau_0 = 0$, $\tau_k = \infty$, $\beta_0 = 1$, and $\beta_i > 0$ for $i = 1, 2, \dots, k-1$. If we take $\lambda(t) = \alpha t^{\alpha-1}/\theta$, i.e., the lifetime under the stress level s_1 has a Weibull distribution with the shape parameter α and the scale parameter θ, then Eq. (1.7) becomes

$$\tilde{\lambda}(t) = \left(\prod_{j=0}^{i-1} \beta_j\right) \alpha t^{\alpha-1}/\theta \quad \text{if } \tau_{i-1} \le t < \tau_i \quad i = 1, \dots, k. \tag{1.8}$$

The CDF associated with the hazard function (1.8) becomes

$$F(t) = 1 - e^{-(t^{\alpha} - \tau_{i-1}^{\alpha})/\theta_i - \sum_{j=1}^{i-1}(\tau_j^{\alpha} - \tau_{j-1}^{\alpha})/\theta_j} \quad \text{if } \tau_{i-1} \le t < \tau_i \quad i = 1, \dots, k,$$

where $\theta_i = \theta \prod_{j=0}^{i-1} \frac{1}{\beta_j}$, for $i = 1, 2, \ldots, k$. This model was also proposed independently by Khamis and Higgins [24] as a power transformed model of the exponential distribution, and it is known as the Khamis and Higgins model (KHM). They have assumed that the lifetime at the stress level s_i has a Weibull distribution with the common shape parameter α but different scale parameter θ_i. Sha and Pan [25] called this model the Weibull proportional hazard model. Xu and Tang [26] showed that the KHM is a special case of the TFRM. It may be recalled that the CDF of a Weibull distribution coincides with the CDF of an exponential distribution under the power transformation, and it is extensively used to prove many properties of a Weibull distribution. The CDF of the CEM of the Weibull distribution does not coincide with the CDF of the CEM of the exponential distribution under the power transformation, whereas the same is true in the case of the KHM.

1.5.4 Cumulative risk model

Note that in all the previous three models, namely the CEM, TRVM, and TFRM, although the CDFs are continuous functions, the corresponding hazard functions are not continuous at the points at which the stress levels are changed. In other words, the effect of the change in the stress level is instantaneous. This is clearly not reasonable for most of the applications. The effect of a change in stress will produce an increase in the risk, but it seems quite likely that one will observe a lag or latency period. The assumption of an instantaneous jump in the hazard function, though unrealistic, leads to a more simple and tractable model.

To overcome this problem Drop et al. [27] considered a more realistic step-stress model that accounts for these latency periods using a piecewise hazard model. Kannan et al. [28] named it the cumulative risk model (CRM). The model can be described as follows. Consider a simple experiment wherein the stress is changed only once during the experiment at the time point τ_1. It is assumed that the hazard function associated with the initial stress level s_1 is $h_1(t)$ and at the stress level s_2 is $h_2(t)$. Both $h_1(t)$ and $h_2(t)$ are continuous functions. The effect of increasing (changing) the stress is not seen immediately, and it is assumed that there is a latency period δ before the effects are completely observed. It is further assumed that in the interval $[\tau_1, \tau_1 + \delta]$, the hazard function increases (changes) slowly. Therefore, the piecewise hazard function of the CRM has the following form:

$$h(t) = \begin{cases} h_1(t) & \text{if} \quad 0 < t < \tau_1 \\ a + bt & \text{if} \quad \tau_1 \le t < \tau_1 + \delta \\ h_2(t) & \text{if} \quad t \ge \tau_1 + \delta. \end{cases}$$

Here the parameters a and b are such that the hazard function $h(t)$ becomes continuous, i.e.,

$$a + b\tau_1 = h_1(\tau_1) \quad \text{and} \quad a + b\tau_2 = h_2(\tau_1 + \delta).$$

Kannan et al. [28] developed the classical inference of the unknown parameters for exponential CRM.

1.6   Organization of the monograph

The rest of the manuscript is organized as follows. In Chapter 2 we consider the CEM and when the lifetime distributions of the experimental units follow different distributions. One-parameter exponential distribution has been considered by different authors since the work of Xiong [29]. Balakrishnan et al. [30] provided the confidence intervals of the unknown parameters based on the exact distributions of the maximum likelihood estimators (MLEs). We further provide the analysis of a simple SSLT model when the lifetime distributions of the experimental units follow two-parameter exponential distribution obtained by Mitra et al. [31]. Several other lifetime distributions, namely Weibull, gamma, log-normal, generalized exponential, Birnbaum-Saunders, Pareto, and geometric, have been considered by several authors; see for example Kateri and Balakrishnan [32], Alkhalfan [33], Alhadeed and Yang [34], Sun and Shi [35], Kamal et al. [36], Arefi and Razmkhah [37], and the references cited therein. Inferential procedures and other related issues will be discussed in detail for different lifetime distributions based on the CEM assumptions. Finally we will discuss the multiple step-stress models.

In Chapter 3 we will be discussing the analysis of the step-stress data based on different model assumptions other than the CEM assumption. First the TRVM which was proposed by DeGroot and Goel [19] will be considered. DeGroot and Goel [19] assumed that the lifetime distributions under different stress levels follow exponential distribution with different scale parameters. Then we will be discussing about the TFRM of Bhattacharyya and Soejoeti [21] and its extension as proposed by Madi [23]. Finally we end up this chapter by discussing different inferential procedures for the different lifetime distributions in the case of CRM.

In reliability analysis it is quite common that more than one risk factor is present at the same time. An investigator is often interested in the assessment of a specific risk in the presence of other risk factors. In this situation the data usually consists of the failure time and an indicator denoting the cause of failure. Two different statistical models are available to analyze such data. They are known as the competing risks model and complementary risks model. Extensive work has been done in the statistical literature discussing different issues related to the competing risks model although not much work has been done on the complementary risks model. Analysis of step stress data based on the competing risks model has been considered by Klein and Basu [38, 39], Pascual [40, 41], Balakrishnan and Han [42], Han and Balakrishnan [43], Liu and Qiu [44], and Han and Kundu [45]. Han [46] considered the analysis of step-stress data based on the complementary risks model. All these will be discussed in detail in Chapter 4.

In Chapter 5 we discuss different miscellaneous topics which we have not mentioned in the previous chapters but they are related to the step stress modeling and its analysis. We provide several references related to step-stress models for further reading. Finally we would like to mention that in each chapter we will be providing several open problems for future work.

Cumulative exposure model

2

2.1 Introduction

The cumulative exposure model (CEM) is the most commonly used model to analyze the lifetime data obtained from a step-stress experiment. Based on the CEM assumption an extensive amount of work has been done on the analysis of lifetime data obtained from a step-stress life test (SSLT) experiment under different censoring schemes. The model was originally proposed by Sediakin [15]. Nelson [16] explained the CEM assumptions explicitly, and provided the maximum likelihood estimation procedures of the unknown parameters when the lifetime distributions follow Weibull and inverse power law distributions. Xiong [29] and Xiong and Milliken [47] considered the classical inference of the unknown parameters for the exponential distribution when the data are Type-II censored. They provided the exact distribution of the maximum likelihood estimators (MLEs). However, the distribution of the MLEs presented there seems to be incorrect. To be more specific, once it is given that $N_1 \geq 1$ failures occurred before τ_1, it is clear that $nt_{1:n}$ cannot exceed $n\tau_1$, and hence clearly it cannot have an exponential distribution as mentioned in Xiong [29]. Balakrishnan et al. [30] first observed this point and they presented the corrected form of the distribution of the MLEs. The exact distributions of the MLEs are a generalized mixture of shifted gamma distributions. Based on the exact distribution of the MLEs, and using the monotonicity properties of the cumulative distribution function (CDF) of the MLEs, as established by Balakrishnan and Iliopoulos [48], the exact confidence intervals of the unknown scale parameters can be obtained by solving two nonlinear equations.

Following the work of Balakrishnan et al. [30], similar results have been obtained for other censoring schemes. For example, Balakrishnan et al. [49] provided the statistical inferences of the scale parameters based on the Type-I censored step-stress data. Balakrishnan and Xie [50, 51] obtained the exact inference of the unknown parameters for a simple step-stress model for a hybrid Type-I and Type-II censored sample. Xie et al. [52] provided the exact inference of the unknown parameters based on progressive Type-II censored step-stress data. In all these cases the exact conditional distribution of the MLEs of the unknown parameters can be obtained and based on these distributions confidence intervals can be constructed. Ganguly et al. [53] provided the Bayesian inference of the unknown scale parameters based on the conjugate gamma priors for different censoring schemes. The Bayes estimates (BEs) under the squared error loss functions cannot be obtained in explicit forms, and the Gibbs sampling procedure has been suggested to compute the credible intervals

Analysis of Step-Stress Models. http://dx.doi.org/10.1016/B978-0-12-809713-7.00002-8

(CRIs) of the unknown parameters. Balakrishnan [54] provided an excellent synthesis of all the exact inferential results for exponential step-stress models based on the classical approach.

Recently Mitra et al. [31] considered the simple step-stress model when the lifetime distributions at the two different stress levels follow a two-parameter exponential distribution with the same location parameter. Based on the CEM assumptions, they obtained the exact distributions of the MLEs. Using the exact distributions of the MLEs, confidence intervals of the unknown parameters can be constructed. They also obtained the BEs of the parameters based on the importance sampling technique under a fairly flexible set of priors. Kateri and Balakrishnan [32] considered the simple step-stress model when the lifetime distributions at the two different stress levels are assumed to follow a two-parameter Weibull distribution with the same shape parameter at the two different stress levels but different scale parameters. Based on the CEM assumptions they obtained the MLEs of the unknown parameters. It is observed that the MLEs cannot be obtained in closed form, and one needs to use some iterative procedure such as the Newton-Raphson method to compute the MLEs. The authors presented some alternative estimators which are easier to compute and they may be used as an initial estimates to compute the MLEs.

Alkhalfan [33] in her PhD thesis considered the simple step-stress model and it is assumed that the lifetime distributions at the two different stress levels follow two-parameter gamma distributions with the same shape parameter, but different scale parameters. She obtained the MLEs of the unknown parameters under different censoring schemes. The MLEs cannot be obtained in explicit forms and numerical techniques are needed to compute the MLEs. Alhadeed and Yang [34] first analyzed the data obtained from a simple SSLT, when the lifetime distributions of the ex-perimental units are assumed to follow log-normal distributions with the same scale parameter but different location parameters at the two different stress levels. Alhadeed and Yang [34] considered the case when the data are not censored. Balakrishnan et al. [55] considered the same problem when the data are Type-I censored. Lin and Chou [56] extended the results for multiple step-stress model. Abdel-Hamida and Al-Hussaini [57] considered the simple step-stress model when it is assumed that the lifetime distributions follow exponentiated exponential or generalized exponential distributions at the two different stress levels with the different shape and scale parameters.

Although extensive work has been done on different inferential issues related to the step-stress models for different continuous lifetime distributions, not much work has been done when the lifetime distributions are discrete. Recently Arefi and Razmkhah [37] and Wang et al. [58] considered the inferential issues of the unknown parameters related to a simple step-stress model when the lifetimes of the experimental units follow geometric distributions at the two different stress levels. They obtained the MLEs of the unknown parameters and provided some inferential results.

The rest of the chapter is organized as follows. In Section 2.2 we consider the one-parameter exponential distribution. We present here in detail the results related to Type-II censoring both for the classical and Bayesian approaches. Since

Balakrishnan [54] provided a synthesis of the results for different censoring schemes for the classical approach, we briefly mention those results for completeness purposes. The details are not provided to avoid repetition. In Section 2.3 we provide the results related to the two-parameter exponential distribution for different censoring schemes. We present the results for the Weibull and generalized exponential distributions in Sections 2.4 and 2.5, respectively. In Section 2.6 we provide the results for several other continuous distributions; viz., gamma, log-normal, log-logistic, Pareto, and Birnbaum-Saunders distributions. The results for geometric distributions are presented in Section 2.7. Finally in Section 2.8 the results related to multiple step-stress models have been presented. In each section we have indicated several open problems for further work.

2.2 One-parameter exponential distribution

The one-parameter exponential distribution plays an important role in statistical reliability theory mainly due to its simple structure and analytical tractability. The one-parameter exponential distribution with mean $\theta > 0$ will be denoted by $\mathrm{Exp}(\theta)$ and it has the following probability density function (PDF):

$$f(x;\theta) = \begin{cases} \frac{1}{\theta}e^{-\frac{x}{\theta}} & \text{if } x > 0 \\ 0 & \text{if } x \leq 0. \end{cases}$$

The PDF of an exponential distribution is a decreasing function on the positive half of the real line, and it has a constant hazard function. Because of its simple analytical structure it is possible to obtain exact inferential results in many different situations. For different properties of an exponential distribution the readers are referred to Johnson et al. [59].

In this section it is assumed that the lifetime distribution of the experimental units follow $\mathrm{Exp}(\theta_j)$ at the stress level s_j, for $j = 1$ and 2. Since it is assumed that the CDF satisfies the CEM assumption, the CDF of the multiple step-stress model can be written as Eq. (1.2). Hence for a simple step-stress model the CDF becomes

$$F(t) = \begin{cases} 0 & \text{if } t < 0 \\ 1 - e^{-\frac{t}{\theta_1}} & \text{if } 0 \leq t < \tau_1 \\ 1 - e^{-\frac{t-\tau_1}{\theta_2}-\frac{\tau_1}{\theta_1}} & \text{if } \tau_1 \leq t < \infty. \end{cases}$$

The corresponding PDF has the following form:

$$f(t) = \begin{cases} 0 & \text{if } t < 0 \\ \frac{1}{\theta_1}e^{-\frac{t}{\theta_1}} & \text{if } 0 \leq t < \tau_1 \\ \frac{1}{\theta_2}e^{-\frac{t-\tau_1}{\theta_2}-\frac{\tau_1}{\theta_1}} & \text{if } \tau_1 \leq t < \infty. \end{cases}$$

Hence, the hazard function of $F(t)$ is

$$h(t) = \begin{cases} \frac{1}{\theta_1} & \text{if} \quad 0 \le t < \tau_1 \\ \frac{1}{\theta_2} & \text{if} \quad \tau_1 \le t < \infty. \end{cases} \tag{2.1}$$

Therefore, from Eq. (2.1) another interpretation can be given for the CEM when the lifetime distributions of the experimental units follow an exponential distribution. In this case the hazard function of an experimental unit remains constant at $1/\theta_1$ at the stress level s_1, and then it becomes $1/\theta_2$ when the stress changes to the level s_2 at τ_1. We will see in the subsequent sections that the same interpretations may not hold for other lifetime distributions. Now we provide the detailed results related to a simple step-stress model when the data are Type-II censored, i.e., it is assumed that the experiment stops when the rth failure takes place. Here N_1 denotes the number of failures before τ_1.

2.2.1 Maximum likelihood estimators

The form of the data is available in Section 1.4. Based on the available data, the likelihood function can be written as follows (see for example Balakrishnan et al. [30] or Balakrishnan [54]):

$$L(\theta_1, \theta_2) = \begin{cases} \frac{c_1}{\theta_2^r} e^{B(\theta_1, \theta_2)} & \text{if} \quad n_1 = 0 \\ \frac{c_2}{\theta_1^{n_1} \theta_2^{r-n_1}} e^{C(\theta_1, \theta_2)} & \text{if} \quad 1 \le n_1 < r \\ \frac{c_1}{\theta_1^r} e^{A(\theta_1)} & \text{if} \quad n_1 = r, \end{cases} \tag{2.2}$$

where

$$c_1 = n!/r!, \quad c_2 = n!/[n_1!(n-n_1)!], \quad A(\theta_1) = -\frac{1}{\theta_1}\left\{ \sum_{i=1}^{r} t_{i:n} + (n-r)t_{r:n} \right\},$$

$$B(\theta_1, \theta_2) = -\frac{1}{\theta_2}\left\{ \sum_{i=1}^{r}\left(\left(\frac{\theta_2}{\theta_1}\right)\tau_1 + t_{i:n} - \tau_1 \right) + (n-r)\left(\left(\frac{\theta_2}{\theta_1}\right)\tau_1 + t_{r:n} - \tau_1 \right) \right\}$$

and

$$C(\theta_1, \theta_2) = -\frac{1}{\theta_1}\left\{ \sum_{i=1}^{n_1} t_{i:n} + (n-n_1)\tau_1 \right\}$$

$$-\frac{1}{\theta_2}\left\{ \sum_{i=1}^{n_1+1} (t_{i:n} - \tau_1) + (n-r)(t_{r:n} - \tau_1) \right\}.$$

From the likelihood function (2.2), the following observations are clear: If $n_1 = r$ or $n_1 = 0$, $\sum_{i=1}^{r} t_{i:n} + (n - r)t_{r:n}$ is a complete sufficient statistic. If $1 \leq n_1 \leq r - 1$, $\left(n_1, \sum_{i=1}^{n_1} t_{i:n}, \sum_{i=n_1+1}^{r} t_{i:n} + (n - r)t_{r:n}\right)$ is a complete sufficient statistic. It is also clear that the MLE of θ_1 does not exist if $n_1 = 0$, and the MLE of θ_2 does not exist if $n_1 = r$. The MLEs of θ_1 and θ_2 exist only when $1 \leq n_1 \leq r - 1$, and they can be obtained by maximizing the corresponding likelihood function. The MLEs of θ_1 and θ_2 are given by

$$\hat{\theta}_1 = \frac{\sum_{i=1}^{n_1} t_{i:n} + (n - n_1)\tau_1}{n_1}, \tag{2.3}$$

$$\hat{\theta}_2 = \frac{\sum_{i=n_1+1}^{r} (t_{i:n} - \tau_1) + (n - r)(t_{r:n} - \tau_1)}{r - n_1}, \tag{2.4}$$

respectively. Clearly, $\hat{\theta}_1$ and $\hat{\theta}_2$ as in Eqs. (2.3), (2.4), respectively, are the conditional MLEs of θ_1 and θ_2, conditional on the event $1 \leq N_1 \leq r - 1$.

It is interesting to see that in this case the MLEs of θ_1 and θ_2 can be obtained in explicit forms. Balakrishnan et al. [30] obtained the conditional distributions of the MLEs, and used the conditional distributions to construct the confidence intervals of the unknown parameters.

2.2.2 Exact conditional distribution of the MLEs

Xiong [29] first presented the exact confidence intervals of θ_1 and θ_2. Balakrishnan et al. [30] pointed out some errors in the derivation. The derivation of the joint PDF of $\hat{\theta}_1$ and $\hat{\theta}_2$ is quite complicated. Balakrishnan et al. [30] provided the exact marginal (conditional) distributions of $\hat{\theta}_1$ and $\hat{\theta}_2$ based on the conditional moment generating function (CMGF). The basic idea of this approach was first suggested by Bartholomew [60]. In this method first the CMGFs of $\hat{\theta}_1$ and $\hat{\theta}_2$ are obtained, and then, inverting them, the corresponding exact conditional distributions of the MLEs can be identified.

We have the following results whose proofs can be obtained in Balakrishnan et al. [30].

Theorem 2.2.1. *The conditional PDFs of $\hat{\theta}_1$ and $\hat{\theta}_2$, conditional on the event $1 \leq N_1 \leq r - 1$, are given by*

$$f_{\hat{\theta}_1}(t) = \sum_{j=1}^{r-1} \sum_{k=0}^{j} c_{j,k} f_G\left(t - \tau_{j,k}; j, \frac{j}{\theta_1}\right) \tag{2.5}$$

and

$$f_{\hat{\theta}_2}(t) = \frac{1}{\sum_{i=1}^{r-1} p_i} \sum_{j=1}^{r-1} p_{r-j} f_G\left(t; j, \frac{j}{\theta_2}\right), \tag{2.6}$$

respectively. Here $f_G(y; \alpha, \lambda)$ is the PDF of a Gamma(α, λ) random variable and it is given by

$$f_G(x; \alpha, \lambda) = \begin{cases} \frac{\lambda^\alpha}{\Gamma(\alpha)} x^{\alpha-1} e^{-\lambda x} & if \quad x \geq 0 \\ 0 & if \quad x < 0, \end{cases}$$

$$c_{j,k} = \frac{(-1)^k}{\sum_{i=1}^{r-1} p_i} \binom{n}{j}\binom{j}{k} e^{-(n-j+k)\frac{\tau_1}{\theta_1}}, \quad and \quad \tau_{j,k} = \frac{\tau_1}{j}(n-j+k), \quad (2.7)$$

$$p_j = \binom{n}{j}\left(1 - e^{-\tau_1/\theta_1}\right)^j e^{-(n-j)\tau_1/\theta_1}.$$

The PDF (2.5) is a generalized mixture of shifted gamma PDFs. In this case some of the mixing coefficients $c_{j,k}$ can be negative. The PDF (2.6) is a mixture of gamma PDFs. Integrating both sides of Eq. (2.5), we obtain the following identity

$$\sum_{j=1}^{r-1} \sum_{k=0}^{j} c_{j,k} = \sum_{j=1}^{r-1} \sum_{k=0}^{j} \frac{(-1)^k}{\sum_{i=1}^{r-1} p_i} \binom{n}{j}\binom{j}{k} e^{-(n-j+k)\frac{\tau_1}{\theta_1}} = 1.$$

From the explicit expressions of $f_{\hat{\theta}_1}(t)$ and $f_{\hat{\theta}_2}(t)$ different properties of $\hat{\theta}_1$ and $\hat{\theta}_2$ can be established. For example, the first two moments of $\hat{\theta}_1$ and $\hat{\theta}_2$ are as follows:

$$E\left(\hat{\theta}_1\right) = \sum_{j=1}^{r-1} \sum_{k=0}^{j} c_{j,k}(\tau_{j,k} + \theta_1) = \theta_1 + \sum_{j=1}^{r-1} \sum_{k=0}^{j} c_{j,k}\tau_{j,k}, \tag{2.8}$$

$$E\left(\hat{\theta}_1^2\right) = \sum_{j=1}^{r-1} \sum_{k=0}^{j} c_{j,k}\left(\frac{(j+1)}{j}\theta_1^2 + \tau_{j,k}^2 + 2\theta_1\tau_{j,k}\right), \tag{2.9}$$

$$E\left(\hat{\theta}_2\right) = \frac{1}{\sum_{i=1}^{r-1} p_i} \sum_{j=1}^{r-1} p_{r-j}\theta_2 = \theta_2, \tag{2.10}$$

$$E\left(\hat{\theta}_2^2\right) = \frac{1}{\sum_{i=1}^{r-1} p_i} \sum_{j=1}^{r-1} p_{r-j}\frac{(j+1)}{j}\theta_2^2. \tag{2.11}$$

From Eqs. (2.8), (2.10) we can immediately observe that $\hat{\theta}_1$ is a biased estimator of θ_1 and $\hat{\theta}_2$ is an unbiased estimator of θ_2. The expressions of the second moments in Eqs. (2.9), (2.11) can be used to calculate the standard errors of the corresponding MLEs. It is also possible to obtain the tail probabilities or the survival functions of $\hat{\theta}_1$ and $\hat{\theta}_2$ by integrating the PDF (2.5) and (2.6), respectively. These probabilities will be used later for constructing the confidence intervals of θ_1 and θ_2. The tail probabilities of $\hat{\theta}_1$ and $\hat{\theta}_2$ are given by

$$P(\hat{\theta}_1 \geq b) = \sum_{j=1}^{r-1} \sum_{k=0}^{j} c_{j,k}\Gamma\left(j, \frac{j}{\theta_1}\langle b - \tau_{j,k}\rangle\right) \quad and$$

$$P(\hat{\theta}_2 \geq b) = \sum_{j=1}^{r-1} \frac{p_{r-j}}{\sum_{i=1}^{r-1} p_i} \times \Gamma\left(j, \frac{jb}{\theta_2}\right),$$

respectively. Here $\langle x \rangle = \max\{x, 0\}$ and for $a > 0, z \geq 0$, $\Gamma(a, z) = \frac{1}{\Gamma(a)} \int_z^\infty t^{a-1} e^{-t} dt$ is the incomplete gamma function.

2.2.3   Different confidence intervals

Balakrishnan et al. [30] proposed three different confidence intervals and compared their performances using extensive simulation experiments. To construct the exact confidence intervals of θ_1 and θ_2, the following assumptions are needed. For any fixed $b > 0$, $P_{\theta_1}(\hat{\theta}_1 \geq b)$ and $P_{\theta_2}(\hat{\theta}_2 \geq b)$ are strictly increasing functions of θ_1 and θ_2, respectively. This assumption guarantees the invertibility of the pivotal quantities; see Theorem 9.2.2 of Casella and Berger [61]. Several authors including Chen and Bhattacharyya [62], Kundu and Basu [63], Childs et al. [8] have used this method to construct exact confidence intervals in different contexts. Using the three monotonicity lemmas of Balakrishnan and Iliopoulos [64], it can be shown that $\hat{\theta}_1$ and $\hat{\theta}_2$ satisfy these monotonicity assumptions. The method of constructing the exact confidence intervals can be briefly described as follows.

Let $c(\theta)$ be a function such that $P_{\theta_1'}(\hat{\theta}_1 \geq c(\theta_1)) = \alpha/2$. For $\theta_1 < \theta_1'$, we have

$$P_{\theta_1'}(\hat{\theta}_1 \geq c(\theta_1')) = P_{\theta_1}(\hat{\theta}_1 \geq c(\theta_1)) < P_{\theta_1'}(\hat{\theta}_1 \geq c(\theta_1)) = \frac{\alpha}{2}, \tag{2.12}$$

where the inequality follows from the assumption that $P_{\theta_1}(\hat{\theta}_1 \geq b)$ is an increasing function of θ_1 for a fixed b. Now, Eq. (2.12) implies that $c(\theta_1) < c(\theta_1')$, which in turn implies that $c(\theta)$ is an increasing function of θ since $\theta_1 < \theta_1'$. Therefore, $c^{-1}(\theta)$ exists uniquely, and is also an increasing function of θ. From Eq. (2.12), it follows that

$$P_{\theta_1}(c^{-1}(\hat{\theta}_1) \leq \theta_1) = 1 - \frac{\alpha}{2}.$$

Therefore, $\theta_{1L} = c^{-1}(\hat{\theta}_1)$ is the lower bound for the $100(1 - \alpha)\%$ confidence interval of θ_1. Similarly, we obtain $\theta_{1U} = d^{-1}(\hat{\theta}_1)$, where $d^{-1}(\cdot)$ is the inverse of the function $d(\cdot)$ obtained as the solution of the equation

$$P_{\theta_1}(\hat{\theta}_1 \leq d(\theta_1)) = \frac{\alpha}{2}.$$

Then $(\theta_{1L}, \theta_{1U})$ is a $100(1 - \alpha)\%$ confidence interval of θ_1. In this case θ_{1L} and θ_{1U} cannot be obtained in closed form and they need to be calculated by solving the following two nonlinear equations

$$\frac{\alpha}{2} = \sum_{j=1}^{r-1} \sum_{k=0}^{j} \frac{(-1)^k}{\sum_{i=1}^{r-1} p_i(\theta_{1L})} \binom{n}{j}\binom{j}{k} e^{-\frac{\tau_1}{\theta_{1L}}(n-j+k)} \Gamma\left(j, \frac{j}{\theta_{1L}}\langle \hat{\theta}_1 - \tau_{j,k} \rangle\right)$$

$$1 - \frac{\alpha}{2} = \sum_{j=1}^{r-1} \sum_{k=0}^{j} \frac{(-1)^k}{\sum_{i=1}^{r-1} p_i(\theta_{1U})} \binom{n}{j}\binom{j}{k} e^{-\frac{\tau_1}{\theta_{1U}}(n-j+k)} \Gamma\left(j, \frac{j}{\theta_{1U}}\langle \hat{\theta}_1 - \tau_{j,k} \rangle\right),$$

where

$$p_j(\theta) = \binom{n}{j} \left(1 - e^{-\frac{\tau_1}{\theta}}\right)^j e^{-\frac{\tau_1}{\theta}(n-j)}.$$

Using a similar argument, a $100(1 - \alpha)\%$ confidence interval of θ_2 say $(\theta_{2L}, \theta_{2U})$ can be obtained as the solutions of the two nonlinear equations

$$\frac{\alpha}{2} = \sum_{j=1}^{r-1} \frac{p_{r-j}}{\sum_{i=1}^{r-1} p_i} \Gamma\left(j, \frac{j\hat{\theta}_2}{\theta_{2L}}\right),$$

$$1 - \frac{\alpha}{2} = \sum_{j=1}^{r-1} \frac{p_{r-k}}{\sum_{i=1}^{r-1} p_i} \Gamma\left(j, \frac{j\hat{\theta}_2}{\theta_{2U}}\right).$$

Note thet these two equations involve θ_1, which can be replaced by $\hat{\theta}_1$ to obtain the confidence limits for θ_2.

Even though in this case the confidence intervals of θ_1 and θ_2 can be obtained based on the exact distributions of their conditional MLEs, they cannot be obtained in explicit forms. They need to be obtained by solving two nonlinear equations.

Now we provide the asymptotic confidence intervals of θ_1 and θ_2 based on the Fisher information matrix. Let $\mathbf{I}(\theta_1, \theta_2) = ((I_{ij}(\theta_1, \theta_2)))$, for $i, j = 1, 2$, denote the Fisher information matrix of θ_1 and θ_2, where

$$I_{ij}(\theta_1, \theta_2) = -E\left[\frac{\partial \ln L(\theta_1, \theta_2)}{\partial \theta_i \partial \theta_j}\right].$$

We have

$$I_{11}(\theta_1, \theta_2) = E\left[-\frac{N_1}{\theta_1^2} + \frac{2U_1}{\theta_1^3}\right],$$

$$I_{21}(\theta_1, \theta_2) = 0 = I_{21}(\theta_1, \theta_2),$$

$$I_{22}(\theta_1, \theta_2) = E\left[-\frac{r - N_1}{\theta_2^2} + \frac{2U_2}{\theta_2^3}\right],$$

where

$$U_1 = \sum_{i=1}^{N_1} t_{i:n} + (n - N_1)\tau_1,$$

$$U_2 = \sum_{i=N_1+1}^{r} (t_{i:n} - \tau) + (n - r)(t_{r:n} - \tau_1).$$

In this case the observed Fisher information matrix can be obtained as

$$\begin{bmatrix} O_{11} & O_{12} \\ O_{21} & O_{22} \end{bmatrix} = \begin{bmatrix} \frac{n_1}{\theta_1^2} & 0 \\ 0 & \frac{r - n_1}{\theta_2^2} \end{bmatrix}.$$

Hence, the asymptotic variances of $\hat{\theta}_1$ and $\hat{\theta}_2$ become

$$V_1 = \frac{\hat{\theta}_1^2}{n_1} \quad \text{and} \quad V_2 = \frac{\hat{\theta}_2^2}{r - n_1},$$

respectively.

The asymptotic distributions of $\frac{\hat{\theta}_1 - E(\theta_1)}{\sqrt{V_1}}$ and $\frac{\hat{\theta}_2 - E(\theta_2)}{\sqrt{V_2}}$, the pivotal quantities, may then be used to construct $100(1 - \alpha)\%$ confidence intervals of θ_1 and θ_2, respectively, and they are as given here:

$$\left[\hat{\theta}_1 - \left(\sum_{j=1}^{r-1} \sum_{k=0}^{j} c_{j,k} \tau_{j,k} \right) \pm z_{1-\frac{\alpha}{2}} \sqrt{V_1} \right] \quad \text{and} \quad \left[\hat{\theta}_2 \pm z_{1-\frac{\alpha}{2}} \sqrt{V_2} \right].$$

Here z_q is the qth upper percentile point of a standard normal distribution. Note that $\sum_{j=1}^{r-1} \sum_{k=0}^{j} c_{j,k} \tau_{j,k}$ is the bias of $\hat{\theta}_1$ while estimating θ_1. Since it involves θ_1, it can be replaced by $\hat{\theta}_1$ in the construction of the confidence interval.

Balakrishnan et al. [30] used the bias-corrected and accelerated (BCA) percentile bootstrap method in this case, as was originally proposed by Efron and Tibshirani [65, pp. 184–188]. The BCA confidence intervals of θ_1 and θ_2 can be obtained using the following algorithm:

Algorithm 2.1

1. Based on the original sample $(t_{1:n}, t_{2:n}, \ldots, t_{r:n})$, and τ_1, obtain $\hat{\theta}_1$ and $\hat{\theta}_2$, the MLEs of θ_1 and θ_2, respectively.
2. Simulate the first r order statistics out of n from a Uniform $(0, 1)$ distribution, say, $U_{1:n}, U_{2:n}, \ldots, U_{r:n}$.
3. Find n_1 such that $U_{n_1:n} \leq 1 - e^{-\frac{\tau_1}{\hat{\theta}_1}} < U_{n_1+1:n}$.
4. For $j \leq n_1$, $T_{j:n} = -\hat{\theta}_1 \ln(1 - U_{j:n})$, and for $j = n_1 + 1, \ldots, r$, $T_{j:n} = -\hat{\theta}_2 \ln(1 - U_{j:n}) + \tau_1 - \frac{\hat{\theta}_2}{\hat{\theta}_1} \tau_1$.
5. Compute the MLEs of θ_1 and θ_2 based on $T_{1:n}, T_{2:n}, \ldots, T_{n_1:n}, T_{n_1+1:n}, \ldots, T_{r:n}$, say $\hat{\theta}_1^{(1)}$ and $\hat{\theta}_2^{(1)}$.
6. Repeat Steps 2–5, B times and obtain $\hat{\theta}_1^{(1)}, \hat{\theta}_2^{(1)}, \hat{\theta}_1^{(2)}, \hat{\theta}_2^{(2)}, \ldots, \hat{\theta}_1^{(B)}, \hat{\theta}_2^{(B)}$.
7. Arrange $\hat{\theta}_1^{(1)}, \hat{\theta}_1^{(2)}, \ldots, \hat{\theta}_1^{(B)}$ in ascending order and obtain $\hat{\theta}_1^{[1]}, \hat{\theta}_1^{[2]}, \ldots, \hat{\theta}_1^{[B]}$. Similarly, arrange $\hat{\theta}_2^{(1)}, \hat{\theta}_2^{(2)}, \ldots, \hat{\theta}_2^{(B)}$ in ascending order and obtain $\hat{\theta}_2^{[1]}, \hat{\theta}_2^{[2]}, \ldots, \hat{\theta}_2^{[B]}$.
8. A two-sided $100(1 - \alpha)\%$ BCA bootstrap confidence interval of θ_i, say $[\theta_{iL}^*, \theta_{iU}^*]$, is then given by, for $i = 1, 2$,

$$\theta_{iL}^* = \hat{\theta}_i^{([B\alpha_{1i}])}, \quad \theta_{iU}^* = \hat{\theta}_i^{([B(1-\alpha_{2i})])},$$

where

$$\alpha_{1i} = \Phi\left(\hat{z}_{0i} + \frac{\hat{z}_{0i} + z_{\frac{\alpha}{2}}}{1 - \hat{a}_i(\hat{z}_{0i} + z_{\frac{\alpha}{2}})}\right) \quad \text{and} \quad \alpha_{2i} = \Phi\left(\hat{z}_{0i} + \frac{\hat{z}_{0i} + z_{1-\frac{\alpha}{2}}}{1 - \hat{a}_i(\hat{z}_{0i} + z_{1-\frac{\alpha}{2}})}\right).$$

Here, $\Phi(\cdot)$ is the CDF of a standard normal random variable and the value of the bias correction $\hat{z}_{0i}$ can be computed as

$$\hat{z}_{0i} = \Phi^{-1}\left(\frac{\text{number of } \hat{\theta}_i^{(j)} < \hat{\theta}_i}{B}\right); \quad i = 1, 2,$$

where, $\Phi^{-1}(\cdot)$ denotes the inverse function of $\Phi(\cdot)$, the acceleration factor $\hat{a}_i$ is

$$\hat{a}_i = \frac{\sum_{j=1}^{r}(\hat{\theta}_{i(\cdot)} - \hat{\theta}_{i(j)})^3}{6\left[\sum_{j=1}^{r}(\hat{\theta}_{i(\cdot)} - \hat{\theta}_{i(j)})^2\right]^{3/2}}; \quad i = 1, 2,$$

where $\hat{\theta}_{i(j)}$ is the MLE of θ_i based on the original sample with the jth observation $t_{j:n}$ deleted (i.e., the jackknife estimate), $j = 1, 2, \ldots, r$ and $\hat{\theta}_{i(\cdot)} = \sum_{j=1}^{r}\hat{\theta}_{i(j)}/r$.

2.2.4 Bayesian inference

So far nobody has considered explicitly the Bayesian inference of a simple step-stress model in the case of one-parameter exponential distributions under the assumption of CEM. Recently, Ganguly et al. [53] considered the Bayesian inference of a simple step-stress model when the lifetimes follow the Weibull distribution, and Mitra et al. [31] considered the Bayesian inference of a simple step-stress model for the two-parameter exponential distribution. The one-parameter exponential distribution can be obtained as a special case from both the cases. The procedures that are developed by Mitra et al. [31] and Ganguly et al. [53] will be explained in the respective sections.

2.2.5 Numerical comparisons and recommendations

Balakrishnan et al. [30] performed extensive simulation experiments to compare the performances of different confidence intervals in terms of the coverage percentages for different sample sizes, parameter values and values of τ_1. Extensive tables are available in Balakrishnan et al. [30]. The coverage percentages of the confidence intervals based on the exact distribution of the MLEs are observed to be quite close to the preassigned nominal level. The coverage percentages of the BCA bootstrap confidence intervals of θ_2 are very close to the nominal level, but the corresponding performances of θ_1 are not very satisfactory, particularly for small or moderate sample sizes. Moreover, the coverage percentages of the asymptotic method are much smaller than the nominal level, for small n. Therefore, although the confidence intervals based on the exact distribution of the MLEs can be obtained by solving two nonlinear equations, it is better to use it particularly when the sample size is small, otherwise the BCA bootstrap method may be used. If the sample size is very large and τ_1 is not too small, then the asymptotic method may be used.

2.2.6 Illustrative example

Xiong [29] simulated a simple step-stress data with $\theta_1 = e^{2.5} \approx 12.182$, $\theta_2 = e^{1.5} \approx 4.482$, $\tau_1 = 5$, $n = 20$, and $r = 16$. The failure times are given as follows:

$$2.10, 3.60, 4.12, 4.34, 5.04, 5.94, 6.68, 7.09, 7.17, 7.49, 7.60, 8.23, 8.24,$$
$$8.25, 8.69, 12.05.$$

In this case $n_1 = 4$, $n_2 = 12$. Balakrishnan et al. [30] obtained the following results: $\hat{\theta}_1 = 23.52$ and $\hat{\theta}_2 = 5.06$. Different confidence intervals are provided in Table 2.1.

Table 2.1 **Different confidence intervals of θ_1 and θ_2 based on the simulated data of Xiong [29]**

Methods	Confidence interval of θ_1		
	90%	**95%**	**99%**
Bootstrap C.I.	(7.78, 34.05)	(6.03, 34.05)	(4.03, 34.05)
Approx C.I.	(0.00, 35.66)	(0.00, 39.36)	(0.00, 46.60)
Exact C.I.	(11.70, 72.95)	(10.35, 94.78)	(8.26, 168.97)
	Confidence interval of θ_2		
	90%	**95%**	**99%**
Bootstrap C.I.	(5.76, 11.43)	(5.51, 12.80)	(4.90, 13.72)
Approx C.I.	(2.66, 7.46)	(2.20, 7.92)	(1.27, 8.82)
Exact C.I.	(3.33, 8.80)	(3.07, 9.86)	(2.64, 12.53)

From Table 2.1 it is observed that the confidence intervals for θ_2 are significantly narrower than those for θ_1. Since the number of failures before τ_1 is very small, the variability in the estimation of θ_1 is very high.

2.2.7 Simple step-stress model: Other censoring schemes

Balakrishnan et al. [49] considered the simple step-stress model for the exponential distribution for Type-I censored samples. In this case the MLEs of θ_1 and θ_2 can be obtained in explicit forms, and the conditional distributions of $\hat{\theta}_1$ and $\hat{\theta}_2$ have been derived. The exact distributions of $\hat{\theta}_1$ and $\hat{\theta}_2$ can be used to construct confidence intervals of θ_1 and θ_2, respectively. Similarly as the Type-II censoring case, bootstrap and asymptotic confidence intervals have also been suggested and compared by extensive simulation experiments.

Along similar lines, Balakrishnan and Xie [50, 51], Balakrishnan et al. [66], and
Xie et al. [52] obtained the exact likelihood inference of θ_1 and θ_2 for a simple
step-stress model when the data are obtained from the exponential distributions, and
they are hybrid Type-II censored, hybrid Type-I censored, unified hybrid censored,
and progressively Type-II censored, respectively. In each case the conditional MLEs
of θ_1 and θ_2 can be obtained in explicit forms, and their exact distributions have been
derived. The exact distributions have been used to construct confidence intervals of the
unknown parameters.

2.3 Two-parameter exponential distribution

In this section we will describe the procedure in case of two-parameter exponential
distribution. The two-parameter exponential distribution with parameters $-\infty < \mu <
\infty$ and $\theta > 0$ will be denoted by $\text{Exp}(\mu, \theta)$, and it has the PDF

$$f(t; \mu, \theta) = \begin{cases} 0 & \text{if } t < \mu \\ \frac{1}{\theta} e^{-\frac{t-\mu}{\theta}} & \text{if } t \geq \mu. \end{cases}$$

The two-parameter exponential distribution is also analytically quite tractable because
of its simple form. Moreover, due to the presence of the location parameter it is more
flexible than the one-parameter exponential distribution.

In this section it is assumed that the lifetime distributions of the experimental
units follow $\text{Exp}(\mu, \theta_j)$ at the stress level s_j, for $j = 1, 2$. One of the possible
justifications of the assumption of a common location parameter is the presence of
an unknown calibration in the equipment used for measuring lifetimes. Based on the
CEM assumption, the CDF of a simple step-stress model becomes

$$F(t) = \begin{cases} 0 & \text{if } t < \mu \\ 1 - e^{-\frac{t-\mu}{\theta_1}} & \text{if } \mu \leq t < \tau_1 \\ 1 - e^{-\frac{t-\tau_1}{\theta_2} - \frac{t-\mu}{\theta_1}} & \text{if } \tau_1 \leq t < \infty, \end{cases} \qquad (2.13)$$

when $\mu < \tau_1$, and for $\mu \geq \tau_1$, the same is given by

$$F(t) = \begin{cases} 0 & \text{if } t < \mu \\ 1 - e^{-\frac{t-\mu}{\theta_2}} & \text{if } t \geq \mu \geq \tau_1. \end{cases}$$

The associated PDFs for $\mu < \tau_1$ and $\mu \geq \tau_1$ are given by

$$f(t) = \begin{cases} 0 & \text{if } t < \mu \\ \frac{1}{\theta_1} e^{-\frac{t-\mu}{\theta_1}} & \text{if } \mu \leq t < \tau_1 \\ \frac{1}{\theta_2} e^{-\frac{t-\tau_1}{\theta_2} - \frac{t-\mu}{\theta_1}} & \text{if } \tau_1 \leq t < \infty, \end{cases}$$

and

$$
f(t) = \begin{cases} 0 & \text{if } t < \mu \\ \frac{1}{\theta_2} e^{-\frac{t-\mu}{\theta_2}} & \text{if } t \geq \mu \geq \tau_1, \end{cases}
$$

respectively. Mitra et al. [31] considered both the classical and Bayesian inference of a simple step-stress model for the two-parameter exponential distribution when the data are Type-II censored.

2.3.1 Likelihood function and MLEs

It is assumed that the data are available from a simple step-stress model and they are Type-II censored. The form of the available data is provided in Section 1.4. First observe that if $\mu > \tau_1$, and for $\theta_1 > 0$, $\theta_2 > 0$, the likelihood of the observed data is given by

$$
L(\mu, \theta_1, \theta_2) = \frac{1}{\theta_2} e^{-\frac{1}{\theta_2} \{ \sum_{j=1}^{r} t_{i:n} + (n-r) t_{r:n} - n\mu \}}.
$$

Hence, the likelihood function does not provide any information about the parameter θ_1. In fact under this assumption the experiment is not a proper simple step-stress experiment. Therefore, it is reasonable to make the assumption that $\mu < \tau_1$.

The likelihood function of the observed data is given by

$$
L(\mu, \theta_1, \theta_2) = \begin{cases} \frac{1}{\theta_2^r} e^{-\frac{n}{\theta_1}\tau_1 + \frac{n}{\theta_1}\mu - \frac{D_2}{\theta_2}} & \text{if } n_1 = 0 \\ \frac{1}{\theta_1^{n_1}\theta_2^{r-n_1}} e^{-\frac{D_1}{\theta_1} - \frac{n}{\theta_1}(t_{1:n}-\mu) - \frac{D_2}{\theta_2}} & \text{if } 1 \leq n_1 \leq r-1 \\ \frac{1}{\theta_1^r} e^{-\frac{D_1}{\theta_1} - \frac{n}{\theta_1}(t_{1:n}-\mu)} & \text{if } n_1 = r, \end{cases}
$$

where $D_1 = \sum_{j=1}^{n_1} t_{j:n} + (n-n_1)m - nt_{1:n}$, $D_2 = \sum_{j=n_1+1}^{r} t_{j:n} + (n-r)t_{r:n} - (n-n_1)m$, and $m = \min\{\tau_1, t_{r:n}\}$. For $n_1 = 0$, and for fixed θ_1 and θ_2, $L(\mu, \theta_1, \theta_2)$ is maximum at $\mu = t_{1:n} > \tau_1$. Now, for $n_1 = 0$,

$$
L(t_{1:n}, \theta_1, \theta_2) = \frac{1}{\theta_2^r} e^{\frac{1}{\theta_1}n(t_{1:n}-\tau_1) - \frac{D_2}{\theta_2}}; \quad \theta_1 > 0, \theta_2 > 0,
$$

which increases as θ_1 decreases. Hence, there exists a path along which $L(\mu, \theta_1, \theta_2)$ increases for $n_1 = 0$, and the MLE of $(\mu, \theta_1, \theta_2)$ does not exist in this case. Similarly, the MLE of $(\mu, \theta_1, \theta_2)$ also does not exist when $n = r$. For $1 \leq n_1 \leq r-1$, the MLE of $(\mu, \theta_1, \theta_2)$ exists, and is given by $(\hat{\mu}, \hat{\theta}_1, \hat{\theta}_2)$, where

$$
\hat{\mu} = t_{1:n}, \quad \hat{\theta}_1 = \frac{D_1}{n_1}, \quad \hat{\theta}_2 = \frac{D_2}{r - n_1}.
$$

Clearly, this is the conditional MLE of $(\mu, \theta_1, \theta_2)$ conditional on the event $1 \leq N_1 \leq r-1$.

2.3.2 Exact conditional distribution of MLEs

The exact conditional distribution of $\hat{\mu}$, conditioning on the event $1 \leq N_1 \leq r - 1$, is the distribution of the first-order statistics of a sample of size n, i.e., $\hat{\mu}|\{1 \leq N_1 \leq r - 1\} \sim \text{Exp}(\mu, \theta_1/n)$. The exact conditional distributions of $\hat{\theta}_1$ and $\hat{\theta}_2$, conditioning on the event $1 \leq N_1 \leq r - 1$, can be obtained by first calculating the respective CMGFs and then inverting them. The explicit derivations can be found in Mitra et al. [31].

Theorem 2.3.1. *The conditional PDF of $\hat{\theta}_1$, conditioning on the event $1 \leq N_1 \leq r - 1$, is given by*

$$f_{\hat{\theta}_1}(t) = c_{10} f_G \left(\tau_{10} - t; 1, \frac{1}{\theta_1(n-1)} \right) - d_{10} f_G \left(-t; 1, \frac{1}{\theta_1(n-1)} \right)$$

$$+ \sum_{i=2}^{r-1} \sum_{j=0}^{i-1} c_{ij} f_1 \left(t - \tau_{ji}; i - 1, \frac{\theta_1}{i}, \frac{(n-j-1)\theta_1}{i(j+1)} \right)$$

$$- \sum_{i=2}^{r-1} \sum_{j=0}^{i-1} d_{ij} f_1 \left(t; i - 1, \frac{\theta_1}{i}, \frac{(n-j-1)\theta_1}{i(j+1)} \right),$$

where

$$d_{ij} = \frac{(-1)^{i-j-1}}{\sum_{k=1}^{r-1} p_k} \binom{n}{i} \binom{i}{j+1} e^{-\frac{n}{\theta_1}(\tau_1 - \mu)},$$

$$c_{ij} = \frac{(-1)^{i-j-1}}{\sum_{k=1}^{r-1} p_k} \binom{n}{i} \binom{i}{j+1} e^{-\frac{n-j-1}{\theta_1}(\tau_1 - \mu)},$$

$$\tau_{ij} = \frac{1}{i}(n - j - 1)(\tau_1 - \mu),$$

$$p_i = \binom{n}{i} \left(1 - e^{-\frac{\tau_1 - \mu}{\theta_1}} \right)^i e^{-(n-i)\frac{\tau_1 - \mu}{\theta_1}},$$

$$f_1(t; \eta, \xi_1, \xi_2) = \frac{e^{t/\xi_2}}{\Gamma(\eta)\xi_2(1 + \xi_1/\xi_2)^\eta} \int_{\max\{0, (1/\xi_1 + 1/\xi_2)t\}}^{\infty} z^{\eta - 1} e^{-z} dz$$

for $t \in (-\infty, \infty)$, and $f_G(t; \xi_1, \xi_2)$ is the same as defined in Eq. (2.7).

Proof. The proof is available in Mitra et al. [31].   □

Theorem 2.3.2. *The conditional PDF of $\hat{\theta}_2$, conditioning on the event $1 \leq N_1 \leq r - 1$, is given by*

$$f_{\hat{\theta}_2}(t) = \sum_{i=1}^{r-1} c_i f_G \left(t, r - i, \frac{r - i}{\theta_2} \right),$$

where $c_i = \frac{p_i}{\sum_{k=1}^{r-1} p_k}$ and $f_G(t; \xi_1, \xi_2)$ is the same as defined in Eq. (2.7).

Proof. The proof is available in Mitra et al. [31].   □

From Theorem 2.3.1 it can be seen that the PDF of $\hat{\theta}_1$ is a generalized mixture of $r(r-1)$ PDFs, and from Theorem 2.3.2 it is observed that the PDF of $\hat{\theta}_2$ is a mixture of $r-1$ gamma PDFs with different shape and scale parameters. Since the shape of the conditional PDF of $\hat{\theta}_1$ as given in Theorem 2.3.1 is difficult to analyze analytically, the plots of the PDF of $\hat{\theta}_1$ when $n = r = 20$, $\mu = 0, \theta_1 = 12, \theta_2 = 4.5$, for different values of τ_1 are provided in Fig. 2.1. The PDF plots of $\hat{\theta}_2$ are provided in Fig. 2.2, for the same set of parameter values. For comparison purposes, the authors have also presented on the same figures the histograms of $\hat{\theta}_1$ and $\hat{\theta}_2$ based on 10,000 replications. It is clear that the true PDFs match very well with the corresponding histograms. It is also clear that, although the PDF of $\hat{\theta}_2$ is a smooth function for all values of τ_1, the PDF of $\hat{\theta}_1$ is not very smooth for small values of τ_1.

The distribution of $\hat{\mu}$ is same as the conditional distribution of the first-order statistic with a sample of size n from the two-parameter exponential distribution, conditioning on the event that there is at least one failure between the time μ and τ_1. Hence,

$$E(\hat{\mu}) = \frac{\mu}{q} + \frac{\theta_1}{n} - \frac{\tau_1(1-q)}{q},$$

where $q = 1 - e^{-n(\tau_1 - \mu)/\theta_1}$ is the probability of getting at least one failure before the time τ_1. As the MLEs do not exist if the number of failures before time τ_1 is zero, one should choose τ_1 so that q is close to one. Using the previous relation, one can have a bias-reduced estimator of μ as

$$\tilde{\mu} = \hat{\mu} - \frac{\hat{\theta}_1}{n}.$$

2.3.3 Different confidence intervals

In order to construct the exact confidence intervals of θ_1 as in the one-parameter case, one needs to assume that $P_{\theta_1}(\hat{\theta}_1 < b)$ is a monotone function of θ_1 for any fixed b. Empirically, it is observed by Mitra et al. [31] that $P_{\theta_1}(\hat{\theta}_1 < b)$ is not a monotone function of θ_1 for a fixed b. Therefore, the construction of the exact confidence intervals becomes difficult. Due to this reason, asymptotic confidence intervals based on the asymptotic distribution of the MLEs and bootstrap confidence intervals have been constructed for θ_1 and θ_2.

2.3.4 Bayesian inference

As the conditional distributions of the MLEs of θ_1 and θ_2 are quite complicated, the Bayesian analysis seems to be a reasonable alternative. Also it is well known that the bootstrap confidence interval does not work well for a threshold parameter μ, but a proper CRI for μ can be obtained in a standard manner. In evaluating the BEs, we mainly consider the squared error loss function, although any other loss function can be easily incorporated in a similar manner.

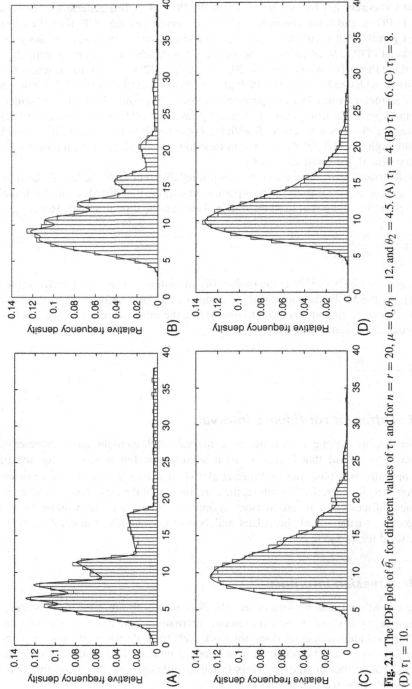

Fig. 2.1 The PDF plot of $\widehat{\theta}_1$ for different values of τ_1 and for $n = r = 20$, $\mu = 0$, $\theta_1 = 12$, and $\theta_2 = 4.5$. (A) $\tau_1 = 4$. (B) $\tau_1 = 6$. (C) $\tau_1 = 8$. (D) $\tau_1 = 10$.

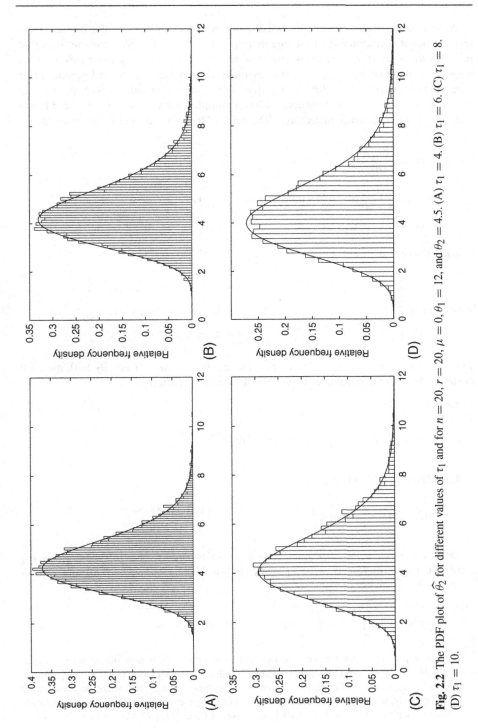

Fig. 2.2 The PDF plot of $\widehat{\theta}_2$ for different values of τ_1 and for $n = 20$, $r = 20$, $\mu = 0$, $\theta_1 = 12$, and $\theta_2 = 4.5$. (A) $\tau_1 = 4$. (B) $\tau_1 = 6$. (C) $\tau_1 = 8$. (D) $\tau_1 = 10$.

Note that the basic aim of the SSLT is to get more failures at the higher stress level, hence it is reasonable to assume that $\theta_1 > \theta_2$. One of the prior assumptions that supports $\theta_1 > \theta_2$ is $\theta_2 = \alpha\theta_1$, where $0 < \alpha < 1$. The following prior assumptions have been made on θ_2, α, and μ. It is assumed that θ_2 has an inverted gamma prior with parameters $a > 0$ and $b > 0$, α has a beta prior distribution with parameters $c > 0$, $d > 0$, the location parameter μ has a noninformative prior over $(-\infty, \tau)$, and they are all independently distributed. The prior PDFs of θ_2, α, and μ are given by

$$\pi_1(\theta_2) = \frac{b^a}{\Gamma(a)} \frac{e^{-b/\theta_2}}{\theta_2^{a+1}}; \quad \theta_2 > 0,$$

$$\pi_2(\alpha) = \frac{1}{B(c,d)} \alpha^{c-1}(1-\alpha)^{d-1}; \quad 0 < \alpha < 1,$$

$$\pi_3(\mu) = 1; \quad -\infty < \mu < \tau_1.$$

The likelihood function becomes

$$L(\mu, \alpha, \theta_2) \propto \begin{cases} \frac{\alpha^{n_1}}{\theta_2^r} e^{-\frac{1}{\theta_2}(\alpha D_3 + D_2 - n\alpha\mu)} & \text{if } \mu < t_{1:n} < \cdots < t_{n_1:n} < \tau_1 < t_{n_1+1:n} < t_{r:n} \\ \frac{\alpha^r}{\theta_2^r} e^{-\frac{1}{\theta_2}(\alpha D_3 - n\alpha\mu)} & \text{if } \mu < t_{1:n} < \cdots < t_{r:n} < \tau_1 \\ \frac{1}{\theta_2^r} e^{-\frac{1}{\theta_2}(n\tau_1\alpha + D_2 - n\alpha\mu)} & \text{if } \mu < \tau_1 < t_{1:n} < \cdots < t_{r:n}, \end{cases}$$

where $1 \le n_1 \le r - 1$, $D_3 = D_1 + nt_{1:n}$. Hence, for $0 < \alpha < 1$ and $\theta_2 > 0$, the joint posterior PDF of the parameters given the data can be written as

Case 1: $n_1 = 0$

$$\pi(\mu, \alpha, \theta_2) \propto \frac{1}{\theta_2^{r+a+1}} \alpha^{c-1}(1-\alpha)^{d-1} e^{-\frac{1}{\theta_2}(n\alpha\tau_1 + D_2 + b - n\alpha\mu)}; \quad \mu < \tau_1.$$

Case 2: $n_1 \in \{1, \ldots, r\}$

$$\pi(\mu, \alpha, \theta_2) \propto \frac{1}{\theta_2^{r+a+1}} \alpha^{n_1+c-1}(1-\alpha)^{d-1} e^{-\frac{1}{\theta_2}(\alpha D_3 + D_2 + b - n\alpha\mu)}; \quad \mu < t_{1:n}.$$

Observe that $\pi(\mu, \alpha, \theta_2)$ is integrable if $r + a - 1 > 0$ and $n_1 + c - 1 > 0$. The BE $\hat{g}(\mu, \alpha, \theta_2)$ of $g(\mu, \alpha, \theta_2)$ with respect to the squared error loss function is the posterior expectation of $g(\mu, \alpha, \theta_2)$, i.e., it can be expressed as

$$\hat{g}(\mu, \alpha, \theta_2) = \int_0^1 \int_0^\infty \int_{-\infty}^{t_{1:n}} g(\mu, \alpha, \theta_2) \pi(\mu, \alpha, \theta_2) d\mu d\theta_2 d\alpha. \tag{2.14}$$

Clearly, Eq. (2.14) cannot be obtained explicitly in general. It is possible to use numerical methods to compute Eq. (2.14); alternatively, Lindley's approximation (see Lindley [67]) can be used to approximate Eq. (2.14). However, the corresponding CRI cannot be obtained by any of these methods. The importance sampling technique can be used to compute simulation consistent BE and the associated CRI in this case. In Case 2, Eq. (2.14) can be written as

$$\hat{g}(\mu,\alpha,\theta_2) = \frac{\int_0^1 \int_0^\infty \int_{-\infty}^{t_{1:n}} g_1(\mu,\alpha,\theta_2) l_1(\alpha) l_2(\theta_2|\alpha) l_3(\mu|\alpha,\theta_2) d\mu d\theta_2 d\alpha}{\int_0^1 \int_0^\infty \int_{-\infty}^{t_{1:n}} g_2(\mu,\alpha,\theta_2) l_1(\alpha) l_2(\theta_2|\alpha) l_3(\mu|\alpha,\theta_2) d\mu d\theta_2 d\alpha},$$

where

$$g_1(\mu,\alpha,\theta_2) = \frac{g(\mu,\alpha,\theta_2)\alpha^{n_1+c-2}(1-\alpha)^{d-1}}{(D_1\alpha+D_2+b)^{r+a-1}},$$

$$g_2(\mu,\alpha,\theta_2) = \frac{\alpha^{n_1+c-2}(1-\alpha)^{d-1}}{(D_1\alpha+D_2+b)^{r+a-1}},$$

$$l_1(\alpha) = 1; \quad 0 < \alpha < 1,$$

$$l_2(\theta_2|\alpha) = \frac{(D_1\alpha+D_2+b)^{r+a-1}}{\Gamma(r+a-1)} \times \frac{e^{-(D_1\alpha+D_2+b)/\theta_2}}{\theta_2^{r+a}}; \quad \theta_2 > 0$$

$$l_3(\mu|\alpha,\theta_3) = \frac{n\alpha}{\theta_2} e^{-\frac{n\alpha}{\theta_2}(t_{1:n}-\mu)}; \quad \mu < t_{1:n}.$$

Note that $l_3(\mu|\alpha,\theta_2)$ has a closed and invertible CDF, hence, it is quite simple to generate samples from this PDF. It also may be mentioned that the preceding choices of g_1, g_2, l_1, l_2, and l_3 functions are not unique, but it is observed by Mitra et al. [31] using extensive simulation experiments that the performances based on above choices are quite satisfactory. The following algorithm can be used to compute a simulation consistent estimate of $\hat{g}(\mu,\alpha,\theta_2)$, and the associated CRI.

Algorithm 2.2
Step 1: Generate α from U(0, 1).
Step 2: Generate θ_2 from IGamma$(r+a-1, D_1\alpha+D_2+b)$.
Step 3: Generate μ from $l_3(\mu|\alpha,\theta_2)$.
Step 4: Repeat steps 1–3, M times to obtain $\{(\mu_1,\alpha_1,\theta_{21}),\ldots,(\mu_M,\alpha_M,\theta_{2M})\}$.
Step 5: Calculate $g_i = g_1(\mu_i,\alpha_i,\theta_{2i})$, for $i = 1,\ldots,M$.
Step 6: Calculate $w_i = g_2(\mu_i,\alpha_i,\theta_{2i})$, for $i = 1,\ldots,M$.
Step 7: Calculate normalizing weight $w_i^* = \dfrac{w_i}{\sum_{j=1}^{M} w_j}$ for $i = 1,\ldots,M$.

Step 8: Approximate Eq. (2.14) by

$$\hat{g}(\mu,\alpha,\theta_2) \approx \sum_{i=1}^{M} w_i^* g_i.$$

Step 9: To find a $100(1-\gamma)\%$, $0 < \gamma < 1$, CRI for $g(\mu,\alpha,\theta_2)$, arrange $\{g_1,\ldots,g_M\}$ to get $\{g_{(1)} < g_{(2)} < \cdots < g_{(M)}\}$. Arrange $\{w_1^*,\ldots,w_M^*\}$ accordingly, to get $\{w_{(1)}^*, w_{(2)}^*,\ldots,w_{(M)}^*\}$. Note that $w_{(i)}^*$'s are not ordered, they are just associated with the $g_{(i)}$'s. A $100(1-\gamma)\%$ CRI is then given by $(g_{(j_1)}, g_{(j_2)})$, where $1 \leq j_1 < j_2 \leq M$ satisfies

$$\sum_{i=j_1}^{j_2} w_{(i)} \leq 1-\gamma < \sum_{i=j_1}^{j_2+1} w_{(i)}. \tag{2.15}$$

The $100(1 - \gamma)\%$ HPD CRI of $g(\mu, \alpha, \theta_2)$ becomes $(g_{(j_1^*)}, g_{(j_2^*)})$, where $1 \leq j_1^* < j_2^* \leq M$, satisfy

$$\sum_{i=j_1^*}^{j_2^*} w_{(i)} \leq 1 - \gamma < \sum_{i=j_1^*}^{j_2^*+1} w_{(i)}, \quad g_{(j_2^*)} - g_{(j_1^*)} \leq g_{(j_2)} - g_{(j_1)},$$

for all j_1 and j_2 satisfying Eq. (2.15).

For Case 1, the following g_1, g_2, l_1, l_2, and l_3 functions can be used.

$$g_1(\mu, \alpha, \theta_2) = \frac{1}{\alpha} g(\mu, \alpha, \theta_2),$$

$$g_2(\mu, \alpha, \theta_2) = \frac{1}{\alpha},$$

$$l_1(\alpha) = \frac{1}{B(c, d)} \alpha^{c-1} (1 - \alpha)^{d-1}; \quad 0 < \alpha < 1,$$

$$l_2(\theta_2 | \alpha) = \frac{(D_2 + b)^{r+a-1} e^{-(D_2+b)/\theta_2}}{\Gamma(r + a - 1)\theta_2^{r+a}}; \quad \theta_2 > 0,$$

$$l_3(\mu | \alpha, \theta_2) = \frac{n\alpha}{\theta_2} e^{-n\alpha(\tau_1 - \mu)/\theta_2}; \quad \mu < \tau_1.$$

Hence, in this case also the BE and the associated CRI for $g(\mu, \alpha, \theta_2)$ can be obtained using the importance sampling technique as in Case 2.

2.3.5 Numerical comparisons

Mitra et al. [31] performed extensive simulation experiments to compare the performances of the bootstrap confidence intervals, the asymptotic confidence intervals and the CRIs. It is observed that the performances of the bootstrap confidence intervals are better than the asymptotic confidence intervals in terms of the coverage percentages, although the average lengths of bootstrap confidence intervals are slightly larger than those of asymptotic confidence intervals. It is also observed that for fixed sample size n and r, as the value of τ_1 increases, the performance of $\hat{\theta}_1$ improves in terms of lower biases and MSEs on the account of availability of more data points, and as expected the performance of $\hat{\theta}_2$ deteriorates, but very marginally. The performances of the BEs based on the noninformative priors are quite satisfactory. The biases and MSEs of the BEs are smaller than those of the MLEs for all the parameters. The average lengths of the HPD CRIs are also smaller than the corresponding average lengths of the bootstrap confidence intervals, and they maintain the coverage percentages. Based on the extensive simulation experiments BEs with noninformative priors are recommended in this case.

2.3.6 Illustrative example

An artificial data set is generated from the CEM (2.13) with $n = 30$, $\mu = 10$, $\theta_1 = e^{2.5} = 12.18$, $\theta_2 = e^{1.5} = 4.48$, and $\tau_1 = 14.5$. The generated data are provided as follows:

10.05	10.59	12.73	12.99	13.71	14.03	14.34	14.53	14.97	15.37
15.43	15.48	15.60	15.76	16.18	16.46	16.86	16.90	17.02	17.36
17.62	18.06	18.31	18.69	18.94	18.95	22.65	22.89	24.51	25.39

To compute the BEs, it is assumed that $a = b = 0$, $c = d = 1$, and based on the preceding hyperparameters, the Bayes estimates and the associated HPD CRIs are obtained based on $M = 8000$. Two different r values have been considered, namely $r = 30$ and $r = 20$. For $r = 30$, the MLEs of μ, θ_1, and θ_2, are 10.05, 17.22, and 3.50, respectively. The corresponding BEs are 9.93, 8.52, and 5.58, respectively. For $r = 20$, the MLEs of μ, θ_1, and θ_2 are 10.05, 17.21, and 3.69, respectively, and the corresponding BEs are 9.89, 11.25, and 7.37, respectively. Biased reduced estimates of μ for both values of r are the same value, 9.48. Different confidence intervals (CIs) and CRIs for μ, θ_1, and θ_2 are presented in Table 2.2. For all the three parameters, it is observed that the symmetric CRIs are very close to the HPD CRIs. The CRIs are significantly smaller than the bootstrap or asymptotic confidence intervals.

2.4 Weibull distribution

We have discussed so far about the analyses of simple step-stress models when the lifetime distributions of the experimental units follow a one-parameter exponential distribution or two-parameter exponential distribution. Although in both the cases it is possible to obtain the exact distributions of the MLEs, it is well known that the exponential distribution has serious limitations in applying it to real data sets mainly due to the fact that it has a constant hazard function and it always has a decreasing PDF. Due to these reasons several more flexible lifetime distributions, for example, the Weibull distribution, gamma distribution, generalized exponential distribution, log-normal distribution, etc. are often used in practice for data analysis purposes. In this section we describe the inferential procedure for the two-parameter Weibull distribution. It is assumed that the two-parameter Weibull distribution with the shape parameter $\alpha > 0$ and the scale parameter $\theta > 0$ has the following PDF and CDF, respectively,

$$f(t; \alpha, \theta) = \begin{cases} 0 & \text{if } t < 0 \\ \frac{\alpha}{\theta^\alpha} t^{\alpha-1} e^{-\left(\frac{t}{\theta}\right)^\alpha} & \text{if } t \geq 0 \end{cases} \quad \text{and} \quad F(t; \alpha, \theta) = \begin{cases} 0 & \text{if } t < 0 \\ 1 - e^{-\left(\frac{t}{\theta}\right)^\alpha} & \text{if } t \geq 0. \end{cases}$$

The two-parameter Weibull distribution is a generalization of the one-parameter exponential distribution. When the shape parameter is one, the two-parameter Weibull distribution becomes the one-parameter exponential distribution. The two-parameter

Table 2.2 **Different confidence intervals and CRIs of μ, θ_1, and θ_2 based on the simulated data**

r	Methods	Confidence interval of μ	
		95%	99%
30	CRI	(9.89, 9.96)	(9.88, 9.97)
	HPD	(9.90, 9.96)	(9.89, 9.97)
20	CRI	(9.84, 9.94)	(9.83, 9.95)
	HPD	(9.84, 9.93)	(9.83, 9.94)
		Confidence interval of θ_1	
		95%	99%
30	ACI	(4.46, 29.98)	(0.45, 33.96)
	BCI	(7.39, 20.70)	(6.27, 23.62)
	CRI	(6.54, 11.10)	(5.86, 12.10)
	HPD	(6.34, 10.76)	(5.70, 11.68)
20	ACI	(4.46, 29.98)	(0.45, 33.96)
	BCI	(7.39, 20.70)	(6.27, 23.62)
	CRI	(8.15, 14.78)	(7.30, 15.52)
	HPD	(8.28, 14.38)	(7.59, 15.55)
		Confidence interval of θ_1	
		95%	99%
30	ACI	(2.07, 4.93)	(1.62, 5.38)
	BCI	(2.24, 5.05)	(2.07, 4.83)
	CRI	(4.29, 7.27)	(3.84, 7.93)
	HPD	(4.15, 7.05)	(3.73, 7.66)
20	ACI	(1.69, 5.70)	(1.05, 6.33)
	BCI	(2.09, 5.27)	(1.90, 5.90)
	CRI	(5.34, 9.69)	(4.78, 10.17)
	HPD	(5.43, 9.73)	(4.97, 10.19)

Weibull distribution is a very flexible model due to the presence of the shape parameter. The PDF of a Weibull distribution can be a decreasing function or a unimodal function depending on whether the shape parameter is less than one or greater than one. The hazard function of a two-parameter Weibull distribution is a decreasing function if the shape parameter is less than one, and an increasing function if the shape parameter is greater than one. For different properties and for different applications in different fields of a two-parameter Weibull distribution, the readers are referred to Johnson et al. [59].

Kateri and Balakrishnan [32] provided the classical inference for a simple step-stress model when the lifetime of the units follow Weibull distributions with the same shape parameter but different scale parameters, and the data are Type-II censored. It is assumed that the lifetime distributions of the experimental units follow a Weibull distribution with the scale parameter $\theta_j > 0$ at the stress level s_j, for $j = 1$ and 2, and the shape parameter $\alpha > 0$ at both the stress levels. The corresponding CDF $F_j(t)$ and PDF $f_j(t)$, for $j = 1$ and 2, and for $t > 0$, are as follows:

$$F_j(t) = 1 - e^{-\left(\frac{t}{\theta_j}\right)^\alpha} \quad \text{and} \quad f_j(t) = \frac{\alpha}{\theta_j^\alpha} t^{\alpha-1} e^{-\left(\frac{t}{\theta_j}\right)^\alpha}.$$

Based on the CEM assumption, the CDF of a test unit under a simple step-stress model is given by

$$F(t) = \begin{cases} F_1(t) & \text{if } 0 < t < \tau_1 \\ F_2(s + t - \tau_1) & \text{if } \tau_1 \le t < \infty, \end{cases}$$

where s is the solution of the equation $F_2(s) = F_1(\tau_1)$; see for example Section 1.5.1 for details. By solving one can easily obtain

$$s = \frac{\theta_2}{\theta_1}\tau_1.$$

Hence, $F(t)$ becomes

$$F(t) = \begin{cases} 1 - e^{-\left(\frac{t}{\theta_1}\right)^\alpha} & \text{if } 0 < t < \tau_1 \\ 1 - e^{-\left(\frac{\tau_1}{\theta_1} + \frac{t}{\theta_2} - \frac{\tau_1}{\theta_2}\right)^\alpha} & \text{if } \tau_1 \le t < \infty, \end{cases}$$

and the associated PDF becomes

$$f(t) = \begin{cases} \frac{\alpha}{\theta_1^\alpha} t^{\alpha-1} e^{-\left(\frac{t}{\theta_1}\right)^\alpha} & \text{if } 0 < t < \tau_1 \\ \frac{\alpha}{\theta_2^\alpha} \left(\frac{\theta_2}{\theta_1}\tau_1 + t - \tau_1\right)^{\alpha-1} e^{-\left(\frac{\tau_1}{\theta_1} + \frac{t}{\theta_2} - \frac{\tau_1}{\theta_2}\right)^\alpha} & \text{if } \tau_1 \le t < \infty. \end{cases} \tag{2.16}$$

Therefore, the hazard function of $F(t)$ can be written as

$$h(t) = \begin{cases} \frac{\alpha}{\theta_1^\alpha} t^{\alpha-1} & \text{if } 0 < t < \tau_1 \\ \frac{\alpha}{\theta_2^\alpha} \left(\frac{\theta_2}{\theta_1}\tau_1 + t - \tau_1\right)^{\alpha-1} & \text{if } \tau_1 \le t < \infty. \end{cases} \tag{2.17}$$

As expected Eq. (2.17) matches with Eq. (2.1) when $\alpha = 1$, but for $\alpha \ne 1$, Eq. (2.17) does not have the same interpretation as the one-parameter exponential model. In all these cases it is observed that, although the CDF of a simple step-stress model for the CEM is a continuous function, the associated hazard function is not. The hazard function has a discontinuity at τ_1.

Based on PDF (2.16), the likelihood function of the observed data

$$\{t_{1:n} < \cdots < t_{n_1:n} < \tau_1 < t_{n_1+1:n} < \cdots < t_{r:n}\} \tag{2.18}$$

can be written as

$$L(\theta_1, \theta_2, \alpha) \propto$$

$$\begin{cases} \Pi_{i=1}^r f_2\left(\frac{\theta_2}{\theta_1}\tau_1 + t_{i:n} - \tau_1\right)\left(1 - F_2\left(\frac{\theta_2}{\theta_1}\tau_1 + t_{i:n} - \tau_1\right)\right)^{n-r} & \text{if } n_1 = 0 \\ \{\Pi_{i=1}^{n_1} f_1(t_{i:n})\}\left\{\Pi_{i=n_1+1}^r f_2\left(\frac{\theta_2}{\theta_1}\tau_1 + t_{i:n} - \tau_1\right)\right\}\left(1 - F_2\left(\frac{\theta_2}{\theta_1}\tau_1 + t_{i:n} - \tau_1\right)\right)^{n-r} & \text{if } n \le n_1 < r \\ \Pi_{i=1}^r f_1(t_{i:n})(1 - F_1(t_{r:n})) & \text{if } n_1 = r. \end{cases}$$

2.4.1 Maximum likelihood estimators

When $n_1 = 0$ or r, it follows that the MLEs of all the unknown parameters do not exist. Hence, we restrict our attention to $1 \le n_1 \le r - 1$. In this case the log-likelihood function without the additive constant can be written as:

$$l(\theta_1, \theta_2, \alpha) = r \ln \alpha - \alpha n_1 \ln \theta_1 - \alpha(r - n_1) \ln \theta_2 + (\alpha - 1) \sum_{i=1}^{n_1} \ln t_{i:n} - \frac{1}{\theta_1^\alpha} \sum_{i=1}^{n_1} t_{i:n}^\alpha$$

$$+ (\alpha - 1) \sum_{i=n_1+1}^r \ln\left(\frac{\theta_2}{\theta_1}\tau_1 + t_{i:n} - \tau_1\right) - \frac{1}{\theta_2^\alpha} \sum_{i=n_1+1}^r \left(\frac{\theta_2}{\theta_1}\tau_1 + t_{i:n} - \tau_1\right)^\alpha$$

$$- \frac{n - r}{\theta_2^\alpha}\left(\frac{\theta_2}{\theta_1}\tau_1 + t_{r:n} - \tau_1\right)^\alpha. \tag{2.19}$$

The MLEs of θ_1, θ_2, and α can be obtained by maximizing Eq. (2.19) with respect to the unknown parameters. The MLEs which are obtained by maximizing the log-likelihood function (2.19) should be viewed as the conditional MLEs under the condition that $1 \le N_1 \le r-1$. The corresponding normal equations can be obtained by differentiating $l(\theta_1, \theta_2, \alpha)$ with respect to $\theta_1, \theta_2, \alpha$, respectively, and equating them to zero. The normal equations are as follows.

$$\dot{l}_{\theta_1} = -\frac{\alpha n_1}{\theta_1} + \frac{\alpha}{\theta_1^{\alpha+1}} \sum_{i=1}^{n_1} t_{i:n}^\alpha - \frac{(\alpha - 1)\theta_2 \tau_1}{\theta_1^2} \times \sum_{i=n_1+1}^r \left(\frac{\theta_2}{\theta_1}\tau_1 + t_{i:n} - \tau_1\right)^{-1} + \frac{\alpha \tau_1}{\theta_1^2 \theta_2^{\alpha-1}}$$

$$\times \left\{ \sum_{i=n_1+1}^r \left(\frac{\theta_2}{\theta_1}\tau_1 + t_{i:n} - \tau_1\right)^{\alpha-1} + (n - r)\left(\frac{\theta_2}{\theta_1}\tau_1 + t_{r:n} - \tau_1\right)^{n-r} \right\}$$

$$= 0 \tag{2.20}$$

$$\dot{l}_{\theta_2} = -\frac{\alpha(r - n_1)}{\theta_2} + \frac{(\alpha - 1)\tau_1}{\theta_1} \times \sum_{i=n_1+1}^r \left(\frac{\theta_2}{\theta_1}\tau_1 + t_{i:n} - \tau_1\right)^{-1}$$

$$+ \frac{\alpha}{\theta_2^{\alpha+1}} \times \left\{ \sum_{i=n_1+1}^r \left(\frac{\theta_2}{\theta_1}\tau_1 + t_{i:n} - \tau_1\right)^\alpha + (n - r)\left(\frac{\theta_2}{\theta_1}\tau_1 + t_{r:n} - \tau_1\right)^\alpha \right\}$$

$$- \frac{\alpha \tau_1}{\theta_1 \theta_2^\alpha} \times \left\{ \sum_{i=n_1+1}^{r} \left(\frac{\theta_2}{\theta_1} \tau_1 + t_{i:n} - \tau_1 \right)^{\alpha-1} + (n-r) \left(\frac{\theta_2}{\theta_1} \tau_1 + t_{r:n} - \tau_1 \right)^{\alpha-1} \right\}$$
$$= 0 \tag{2.21}$$

$$\dot{l}_\alpha = \frac{r}{\alpha} - n_1 \ln \theta_1 - (r - n_1) \ln \theta_2 + \sum_{i=1}^{n_1} \ln t_{i:n} + \sum_{i=n_1+1}^{r} \ln \left(\frac{\theta_2}{\theta_1} \tau_1 + t_{i:n} - \tau_1 \right)$$

$$+ \frac{1}{\theta_1^\alpha} \left\{ \ln \theta_1 \sum_{i=1}^{n_1} t_{i:n}^\alpha - \sum_{i=1}^{n_1} t_{i:n}^\alpha \ln t_{i:n} \right\} + \frac{1}{\theta_2^\alpha} \left\{ \ln \theta_2 \sum_{i=n_1+1}^{r} \left(\frac{\theta_2}{\theta_1} \tau_1 + t_{i:n} - \tau_1 \right)^\alpha \right.$$

$$\left. - \sum_{i=n_1+1}^{r} \left(\frac{\theta_2}{\theta_1} \tau_1 + t_{i:n} - \tau_1 \right)^\alpha \times \ln \left(\frac{\theta_2}{\theta_1} \tau_1 + t_{i:n} - \tau_1 \right) \right\}$$

$$+ \frac{n-r}{\theta_2^\alpha} \left\{ \ln \theta_2 \left(\frac{\theta_2}{\theta_1} \tau_1 + t_{r:n} - \tau_1 \right)^\alpha \right.$$

$$\left. - \left(\frac{\theta_2}{\theta_1} \tau_1 + t_{r:n} - \tau_1 \right)^\alpha \times \ln \left(\frac{\theta_2}{\theta_1} \tau_1 + t_{r:n} - \tau_1 \right) \right\} = 0. \tag{2.22}$$

The normal equations (2.20)–(2.22) cannot be solved explicitly. One needs to use a numerical procedure like the Newton-Raphson algorithm to solve these nonlinear equations. Alternatively, some nonlinear optimization procedure like the genetic algorithm (see for example Goldberg [68]) or simulated annealing (see Kirkpatrick et al. [69]) may be used for this purpose.

2.4.2 Newton-Raphson algorithm and initial values

In this section we discuss how the Newton-Raphson algorithm can be used in practice for this case. This algorithm requires the inverse of the observed Fisher Information Matrix, $\mathbf{I}_{obs}(\boldsymbol{\theta})$, where $\boldsymbol{\theta} = (\theta_1, \theta_2, \alpha)$ and

$$\mathbf{I}_{obs}(\boldsymbol{\theta}) = - \begin{bmatrix} \frac{\partial^2 l}{\partial \theta_1^2} & \frac{\partial^2 l}{\partial \theta_1 \partial \theta_2} & \frac{\partial^2 l}{\partial \theta_1 \partial \alpha} \\ \frac{\partial^2 l}{\partial \theta_2 \partial \theta_1} & \frac{\partial^2 l}{\partial \theta_2^2} & \frac{\partial^2 l}{\partial \theta_2 \partial \alpha} \\ \frac{\partial^2 l}{\partial \alpha \partial \theta_1} & \frac{\partial^2 l}{\partial \alpha \partial \theta_2} & \frac{\partial^2 l}{\partial \alpha^2} \end{bmatrix} = \begin{bmatrix} O_{11} & O_{12} & O_{13} \\ O_{21} & O_{22} & O_{23} \\ O_{31} & O_{32} & O_{33} \end{bmatrix} \quad \text{(say)}. \tag{2.23}$$

The explicit expressions have been obtained by Kateri and Balakrishnan [32] and they are provided as follows:

$$O_{11} = - \frac{\alpha n_1}{\theta_1^2} - \frac{2(\alpha - 1)\tau_1 \theta_2}{\theta_1^3} F_{11} + \frac{\alpha(\alpha + 1)}{\theta_1^{\alpha+2}} F_{12} + \frac{(\alpha - 1)\tau_1^2 \theta_2^2}{\theta_1^4} F_{19}$$

$$+ \frac{2\alpha \tau_1}{\theta_1^3 \theta_2^\alpha - 1} (F_{13} + F_{14}) + \frac{\alpha(\alpha - 1)\tau_1^2}{\theta_1^4 \theta_2^{\alpha-2}} (F_{1.10} + F_{1.11}).$$

$$O_{22} = -\frac{\alpha(r - n_1)}{\theta_2^2} + \frac{(\alpha - 1)\tau_1^2}{\theta_1^2}F_{19} - \frac{\alpha(\alpha + 1)}{\theta_2^{\alpha+2}}(F_{15} + F_{16}) - \frac{2\alpha^2\tau_1}{\theta_1\theta_2^{\alpha+1}}(F_{13} + F_{14})$$

$$+ \frac{\alpha(\alpha - 1)\tau_1^2}{\theta_1^2\theta_2^\alpha}(F_{1.10} + F_{1.11}).$$

$$O_{12} = O_{21} = \frac{(\alpha - 1)\tau_1}{\theta_1^2}F_{11} - \frac{(\alpha - 1)\theta_2\tau_1^2}{\theta_1^3}F_{19} + \frac{\alpha(\alpha - 1)\tau_1}{\theta_1^2\theta_2^\alpha}(F_{13} + F_{14})$$

$$- \frac{\alpha(\alpha - 1)\tau_1^2}{\theta_1^3\theta_2^{\alpha-1}}(F_{1.10} + F_{1.11}).$$

$$O_{23} = O_{32} = \frac{r - n_1}{\theta_2} - \frac{\tau_1}{\theta_1}F_{11} - \frac{\alpha}{\theta_2^{\alpha+1}}(F_{32} + F_{33}) + \frac{\tau_1\alpha}{\theta_1\theta_2^\alpha}(F_{34} + F_{35})$$

$$- \frac{(1 - \alpha \ln \theta_2)}{\theta_2^{\alpha+1}}(F_{15} + F_{15}) - \frac{\tau_1(1 - \alpha \ln \theta_2)}{\theta_1\theta_2^\alpha}(F_{13} + F_{14}).$$

where

$$F_{11} = \sum_{i=n_1+1}^{r}\left(\frac{\theta_2}{\theta_1}\tau_1 + t_{i:n} - \tau_1\right)^{-1}, \quad F_{12} = \sum_{i=1}^{n_1}t_{i:n}^\alpha,$$

$$F_{13} = \sum_{i=n_1+1}^{r}\left(\frac{\theta_2}{\theta_1}\tau_1 + t_{i:n} - \tau_1\right)^{\alpha-1},$$

$$F_{14} = (n - r)\left(\frac{\theta_2}{\theta_1}\tau_1 + t_{r:n} - \tau_1\right)^{\alpha-1}, \quad F_{15} = \sum_{i=n_1+1}^{r}\left(\frac{\theta_2}{\theta_1}\tau_1 + t_{i:n} - \tau_1\right)^\alpha,$$

$$F_{16} = (n - r)\left(\frac{\theta_2}{\theta_1}\tau_1 + t_{r:n} - \tau_1\right)^\alpha, \quad F_{19} = \sum_{i=n_1+1}^{r}\left(\frac{\theta_2}{\theta_1}\tau_1 + t_{i:n} - \tau_1\right)^{-2},$$

$$F_{1.10} = \sum_{i=n_1+1}^{r}\left(\frac{\theta_2}{\theta_1}\tau_1 + t_{i:n} - \tau_1\right)^{\alpha-2}, \quad F_{1.11} = (n - r)\left(\frac{\theta_2}{\theta_1}\tau_1 + t_{r:n} - \tau_1\right)^{\alpha-2},$$

$$F_{31} = \sum_{i=1}^{n_1}t_{i:n}^\alpha \ln t_{i:n}, \quad F_{32} = \sum_{i=n_1+1}^{r}t_{i:n}^\alpha \ln t_{i:n}, \quad F_{33} = (n - r)t_{r:n}^\alpha \ln t_{r:n},$$

$$F_{34} = \sum_{i=n_1+1}^{r}t_{i:n}^{\alpha-1} \ln t_{i:n}, \quad F_{35} = (n - r)t_{r:n}^{\alpha-1} \ln t_{r:n},$$

$$F_{21} = -\theta_1^{-\alpha}\left((\ln \theta_1)^2 F_{12} - 2F_{31} \ln \theta_1 + \sum_{i=1}^{n_1}t_{i:n}^\alpha(\ln t_{in}^\alpha)^2\right),$$

$$F_{22} = -\theta_2^{-\alpha}\left((\ln \theta_2)^2 F_{15} - 2F_{32} \ln \theta_2 + \sum_{i=n_1+1}^{r}\left(\frac{\theta_2}{\theta_1}\tau_1 + t_{i:n} - \tau_1\right)^\alpha\right.$$

$$\times \left[\ln \left(\frac{\theta_2}{\theta_1} \tau_1 + t_{i:n} - \tau_1 \right) \right]^2 \right).$$

$$F_{23} = -(n-r)\theta_2^{-\alpha}(\ln \theta_2)^2 \left(\frac{\theta_1}{\theta_1} \tau_1 + t_{r:n} - \tau_1 \right)^{\alpha}$$

$$+ 2(n-r)\theta_2^{-\alpha} \left(\frac{\theta_1}{\theta_1} \tau_1 + t_{r:n} - \tau_1 \right)^{\alpha} (\ln \theta_2) \ln \left(\frac{\theta_1}{\theta_1} \tau_1 + t_{r:n} - \tau_1 \right)$$

$$- (n-r)\theta_2^{-\alpha} \left(\frac{\theta_1}{\theta_1} \tau_1 + t_{r:n} - \tau_1 \right)^{\alpha} \left(\ln \left(\frac{\theta_1}{\theta_1} \tau_1 + t_{r:n} - \tau_1 \right) \right)^2.$$

Suppose at the jth iterate the value of $\boldsymbol{\theta}$ is denoted by $\boldsymbol{\theta}^{(j)} = (\theta_1^{(j)}, \theta_2^{(j)}, \alpha^{(j)})$. Then the updating equations of the Newton-Raphson algorithm are

$$\boldsymbol{\theta}^{(j+1)} = \boldsymbol{\theta}^{(j)} + (\boldsymbol{I}_{obs}(\boldsymbol{\theta}^{(j)}))^{-1} \mathbf{u}^{(j)}.$$

Here $\boldsymbol{I}_{obs}$ is the same as defined in Eq. (2.23), and $\mathbf{u}^{(j)} = (\dot{l}_{\theta_1}, \dot{l}_{\theta_2}, \dot{l}_{\alpha})^T$, evaluated at $\boldsymbol{\theta}^{(j)}$ for $j = 0, 1, \ldots$, with $\boldsymbol{\theta}^{(0)}$ being the vector of initial values.

Hence, the question arises how to choose these initial values? The natural choice, which may not be the best one, is start with the exponential case, i.e.,

$$\alpha^{(0)} = 1, \quad \theta_1^{(0)} = \frac{\sum_{i=1}^{n_1} t_{i:n} + (n - n_1)\tau_1}{n_1}$$

$$\theta_2^{(0)} = \frac{\sum_{i=n_1+1}^{r} (t_{i:n} - \tau_1) + (n - r)(t_{r:n} - \tau_1)}{r - n_1}.$$

Alternatively, take some discrete choices of α, say for example, 0.5, 0.75, 1.0, 1.5, 2.0, 2.5, 3.0, etc. and for each α value make a contour plot of the likelihood function $L(\theta_1, \theta_2, \alpha)$ as a function of θ_1 and θ_2. Using the values of α and the contour plot one can obtain initial values of $\alpha^{(0)}$, $\theta_1^{(0)}$, and $\theta_2^{(0)}$.

2.4.3 Confidence intervals

It has been observed that the MLEs cannot be obtained in closed form. Hence, it is not possible to obtain the exact distributions of the MLEs. Therefore, construction of exact confidence intervals of the unknown parameters is not possible. Due to this reason, Kateri and Balakrishnan [32] proposed two different methods to construct confidence intervals, namely; (i) the asymptotic confidence interval when the sample size is large and (ii) the bootstrap confidence interval when the sample size is small or moderate.

When the sample size is large, the asymptotic confidence intervals of the unknown parameters α, θ_1, and θ_2 can be constructed based on the pivotal quantities $(\hat{\alpha} - E(\hat{\alpha}))/\sqrt{V(\hat{\alpha})}$, $(\hat{\theta}_1 - E(\hat{\theta}_1))/\sqrt{V(\hat{\theta}_1)}$, $(\hat{\theta}_2 - E(\hat{\theta}_2))/\sqrt{V(\hat{\theta}_2)}$, respectively. Now using the asymptotic normality of the MLEs, asymptotic two-sided $100(1 - \alpha)\%$ confidence intervals of α, θ_1, and θ_2 can be obtained as

$$\hat{\alpha} \mp z_{\alpha/2}\sqrt{V(\hat{\alpha})}, \quad \hat{\theta}_1 \mp z_{\alpha/2}\sqrt{V(\hat{\theta}_1)}, \quad \hat{\theta}_2 \mp z_{\alpha/2}\sqrt{V(\hat{\theta}_2)},$$

respectively. Here, $V(\hat{\theta}_1)$, $V(\hat{\theta}_2)$, and $V(\hat{\alpha})$ can be obtained as the diagonal elements of the inverse of $\mathbf{I}_{obs}$ as defined in Eq. (2.23). Kateri and Balakrishnan [32] used the percentile bootstrap and the bias corrected bootstrap methods to construct confidence intervals of the unknown parameters.

2.4.4 Numerical comparisons and recommendations

Kateri and Balakrishnan [32] performed extensive simulation experiments to observe the performances of the MLEs mainly in terms of biases and MSEs and to compare the two different confidence intervals based on their coverage percentages and average lengths. It is observed that the performances of the MLEs are quite satisfactory in the sense that as the sample size increases the biases and the MSEs decrease. It indicates the consistency properties of the MLEs. Regarding the confidence intervals it is observed that for both the methods, as the sample size increases, the average lengths decrease and the coverage percentages converge to the corresponding nominal level. For small sample sizes the asymptotic confidence intervals are too conservative, resulting in low coverage percentages. In the case of BCA bootstrap confidence intervals, it is observed that for small or moderate sample sizes, mainly for $n < 50$, the bias correction does not help much, hence it is not recommended. For small or moderate sample sizes the simple percentile bootstrap method works better than the BCA bootstrap method. For large sample sizes BCA bootstrap confidence intervals perform quite well. It is also observed that the coverage percentages of the confidence interval of α are quite close to the corresponding nominal level, whereas those of θ_2 are quite poor and they are becoming worse as the proportion of censoring increases. Overall it is observed that BCA performs the best if the sample size is large and the proportion of censoring is not very high. If the sample size is small or moderate and the proportion of censoring is high, it is better to use the percentile bootstrap method.

It is also observed from the extensive simulation study that for small sample sizes the asymptotic confidence intervals provide very small coverage percentages. This is mainly due to the fact that for small sample sizes the distributions of the MLEs are very skewed and the biases are also quite significant. It results that the asymptotic confidence interval is based on a standard normal distribution centered at the wrong mean. Therefore, if some proper bias correction technique is adopted it might provide better results. More work is needed along this direction.

2.4.5 Illustrative example

The following data set was simulated by Balakrishnan and Xie [50] with $n = r = 35$, $\theta_1 = 25.78, \theta_2 = 12.18, \alpha = 1$, and $\tau_1 = 15$. The data are presented in Table 2.3. Kateri and Balakrishnan [32] provided the MLEs of the unknown parameters and the associated confidence intervals. The MLEs of α, θ_1, and θ_2 are $\hat{\alpha} = 0.943$, $\hat{\theta}_1 = 26.277$, and $\hat{\theta}_2 = 13.810$, respectively. Different confidence intervals are presented in

Table 2.3 **Simulated data set of Balakrishnan and Xie [50]**

$t_{i:n} < \tau$	0.22	0.35	1.27	1.67	2.22	3.79	5.78	8.43
	9.27	10.34	11.85	12.63	12.68	12.85	12.88	13.14
	15.28	16.23	17.21	18.52	19.12	19.39	19.81	22.06
$t_{i:n} > \tau$	23.85	28.46	28.65	28.97	30.02	31.42	35.45	36.25
	57.40	58.46	115.14					

Table 2.4 **Different confidence intervals of θ_1, θ_2, and α based on the simulated data of Balakrishnan and Xie [50]**

Methods	Confidence interval of θ_1		
	90%	95%	99%
Bootstrap C.I.	(19.224, 47.331)	(18.791, 54.970)	(16.439, 74.667)
Asymptotic C.I.	(12.992, 39.562)	(10.446, 42.108)	(5.472, 47.082)
BCA C.I.	(19.121, 44.344)	(18.487, 52.096)	(14.795, 68.211)
	Confidence interval of θ_2		
	90%	95%	99%
Bootstrap C.I.	(8.742, 26.251)	(8.337, 28.855)	(7.033, 40.163)
Asymptotic C.I.	(5.849, 22.131)	(3.895, 23.725)	(0.779, 26.841)
BCA C.I.	(7.311, 21.386)	(6.262, 23.768)	(6.177, 27.601)
	Confidence interval of α		
	90%	95%	99%
Bootstrap C.I.	(0.698, 1.449)	(0.660, 1.606)	(0.576, 1.846)
Asymptotic C.I.	(0.589, 1.297)	(0.522, 1.364)	(0.389, 1.497)
BCA C.I.	(0.623, 1.265)	(0.578, 1.389)	(0.319, 1.622)

Table 2.4. It is clear from Table 2.4 that the percentile bootstrap and BCA bootstrap confidence intervals are very similar in nature, where asymptotic confidence intervals are quite different from these two confidence intervals. Moreover, the confidence lengths for α are significantly smaller than those of θ_1 and θ_2.

2.4.6 Open problems

In this section we mention some open problems related to the Weibull distribution step-stress model for future work.

Open Problem: Kateri and Balakrishnan [32] provided a complete classical inference of the unknown parameters for the Weibull model in the case of Type-II censoring. It will be interesting to develop similar methods for other censoring schemes.

Open Problem: Kateri and Balakrishnan [32] provided the classical inference for the Weibull model; it is important to develop the Bayesian inference and compare their performances. It might be possible to develop a general method for different censoring schemes in this case.

Open Problem: Note that Kateri and Balakrishnan [32] developed the inference based on the assumption that the lifetime of the experimental units follow Weibull distributions with the same shape parameter and different scale parameters at the two different stress levels. The assumption of the equal shape parameter is mainly due to analytical convenience. It is important to develop both classical and Bayesian inference when the two shape parameters are not equal.

Open Problem: Mitra et al. [31] developed classical and Bayesian inference procedures for a simple step-stress model when the lifetime of the experimental units follow a two-parameter exponential distribution. It will be important to generalize that procedure in the case of the Weibull distribution when the location parameter is also present.

Open Problem: It is important to develop both the classical and Bayesian inference for the multiple step-stress model when the lifetime distributions follow Weibull distributions.

2.5 Generalized exponential distribution

The generalized exponential distribution or exponentiated exponential distribution has received considerable attention recently since its introduction by Gupta and Kundu [70]. A two-parameter generalized exponential distribution with the shape parameter $\alpha > 0$ and scale parameter $\lambda > 0$ has the following PDF and CDF, respectively,

$$f(t; \alpha, \lambda) = \begin{cases} 0 & \text{if } t < 0 \\ \alpha \lambda e^{-\lambda t}(1 - e^{-\lambda t})^{\alpha-1} & \text{if } t \geq 0, \end{cases}$$

and

$$F(t; \alpha, \lambda) = \begin{cases} 0 & \text{if } t < 0 \\ (1 - e^{-\lambda t})^{\alpha} & \text{if } t > 0. \end{cases}$$

From the CDF or from the PDF it can be seen that when $\alpha = 1$, it becomes an exponential distribution. Therefore, all the three distributions, namely Weibull, gamma and the generalized exponential distributions, are extensions of exponential distribution but in different ways. It is also a very flexible model and its PDF and the hazard function can take various shapes. If the shape parameter is less than one, the PDF is a decreasing function and if the shape parameter is greater than one, it is unimodal. Similarly, the hazard function is also a decreasing function if the shape parameter is less than one, and it is an increasing function if the shape parameter is greater than one.

The two-parameter generalized exponential distribution is a special case of the three-parameter exponentiated Weibull distribution proposed by Mudholkar and Srivastava [71]; see also Mudholkar et al. [72] in this respect. The generalized exponential distribution was proposed by Gupta and Kundu [70] to analyze lifetime data as an alternative to the well-known gamma or Weibull distribution. In fact it has been observed that if all these three distributions are fitted to a given data set, then it is very difficult to discriminate between these three distributions, particularly if the sample size is not large; see for example Gupta and Kundu [73, 74]. Extensive work has been done in the last 15 years dealing with different aspects of the generalized exponential distribution; interested readers are referred to the review articles by Gupta and Kundu [75], Nadarajah [76] or the recent monograph by Al-Hussaini and Ahsanullah [77].

Abdel-Hamid and Al-Hussaini [57] provided the classical inference of a simple step-stress model when the lifetime distributions of the experimental units follow a generalized exponential distribution having different scale parameters but with the same shape parameter at the two stress levels. Recently, Samanta and Kundu [78] provided the Bayesian inference of the same problem and compared their performances using extensive simulation experiments.

Let us assume that at the stress level s_j, the lifetime distribution of the experimental units follow a generalized exponential distribution with the shape parameter α and the scale parameter λ_j, for $j = 1$ and 2. Therefore, based on the CEM assumptions, the CDF of a test unit under such a simple step-stress model is given by

$$
F(t) = \begin{cases} 0 & \text{if } t < 0 \\ (1 - e^{-\lambda_1 t})^\alpha & \text{if } 0 \le t < \tau_1 \\ (1 - e^{-\lambda_2(t + \frac{\lambda_1}{\lambda_2}\tau_1 - \tau_1)})^\alpha & \text{if } \tau_1 \le t < \infty. \end{cases}
$$

The corresponding PDF becomes

$$
f(t) = \begin{cases} 0 & \text{if } t < 0 \\ \alpha\lambda_1(1 - e^{-\lambda_1 t})^{\alpha-1} e^{-\lambda_1 t} & \text{if } 0 \le t < \tau_1 \\ \alpha\lambda_2(1 - e^{-\lambda_2(t + \frac{\lambda_1}{\lambda_2}\tau_1 - \tau_1)})^{\alpha-1} e^{-\lambda_2(t + \frac{\lambda_1}{\lambda_2}\tau_1 - \tau_1)} & \text{if } \tau_1 \le t < \infty. \end{cases}
$$

Hence, the hazard function of a test unit under such a simple step-stress model can be written as

$$
h(t) = \begin{cases} \dfrac{\alpha\lambda_1(1 - e^{-\lambda_1 t})^{\alpha-1} e^{-\lambda_1 t}}{1 - (1 - e^{-\lambda_1 t})^\alpha} & \text{if } 0 < t < \tau_1 \\[2em] \dfrac{\alpha\lambda_2(1 - e^{-\lambda_2(t + \frac{\lambda_1}{\lambda_2}\tau_1 - \tau_1)})^{\alpha-1} e^{-\lambda_2(t + \frac{\lambda_1}{\lambda_2}\tau_1 - \tau_1)}}{1 - (1 - e^{-\lambda_2(t + \frac{\lambda_1}{\lambda_2}\tau_1 - \tau_1)})^\alpha} & \text{if } \tau_1 \le t < \infty. \end{cases}
$$

In this case also it may be observed that when $\alpha = 1$, the hazard function remains constant at the two different stress levels. Based on the complete data set

$$
\mathcal{D} = \{t_{1:n} < t_{2:n} < \cdots < t_{n_1:n} < \tau_1 < t_{n_1+1:n} < \cdots < t_{n:n}\}, \tag{2.24}
$$

the likelihood function can be written as

$$
L(\alpha, \lambda_1, \lambda_2) \propto \alpha^n \lambda_1^{n_1} \lambda_2^{n-n_1} \prod_{i=1}^{n_1} (1 - e^{-\lambda_1 t_{i:n}})^{\alpha-1} e^{-\lambda_1 \sum_{i=1}^{n_1} t_{i:n}}
$$

$$
\times \prod_{i=n_1+1}^{n} (1 - e^{-\lambda_2(t_{i:n} + \frac{\lambda_1}{\lambda_2}\tau_1 - \tau_1)})^{\alpha-1} e^{-\lambda_2 \sum_{i=n_1+1}^{n} (t_{i:n} + \frac{\lambda_1}{\lambda_2}\tau_1 - \tau_1)}.
$$

Here $0 \leq n_1 \leq n$, and the standard conventions $\prod_{i=1}^{0} = 1$ and $\prod_{i=n+1}^{n} = 1$ are assumed. Now we derive the MLEs of the unknown parameters based on the assumption that $1 \leq n_1 \leq n-1$; otherwise, clearly the MLEs of all the three parameters do not exist.

2.5.1 Maximum likelihood estimators

The maximum likelihood estimators of the unknown parameters can be obtained by maximizing the following log-likelihood function

$$l(\alpha, \lambda_1, \lambda_2) = n \ln \alpha + n_1 \ln \lambda_1 + (n - n_1) \ln \lambda_2 - \lambda_1 \sum_{i=1}^{n_1} t_{i:n}$$

$$+ (\alpha - 1) \sum_{i=1}^{n_1} \ln(1 - e^{-\lambda_1 t_{i:n}}) - \lambda_2 \sum_{i=n_1+1}^{n} \left(t_{i:n} + \frac{\lambda_1}{\lambda_2} \tau_1 - \tau_1 \right)$$

$$+ (\alpha - 1) \sum_{i=n_1+1}^{n} \ln \left(1 - e^{-\lambda_2 \left(t_{i:n} + \frac{\lambda_1}{\lambda_2} \tau_1 - \tau_1 \right)} \right). \tag{2.25}$$

The conditional MLEs of α, λ_1, and λ_2 conditioning on $\{1 \leq N_1 \leq n - 1\}$ can be obtained by solving the following three normal equations:

$$\dot{l}_\alpha = \frac{n}{\alpha} + \sum_{i=1}^{n_1} \ln(1 - e^{-\lambda_1 t_{i:n}}) + \sum_{i=n_1+1}^{n} \ln \left(1 - e^{-\lambda_2 \left(t_{i:n} + \frac{\lambda_1}{\lambda_2} \tau_1 - \tau_1 \right)} \right) = 0, \tag{2.26}$$

$$\dot{l}_{\lambda_1} = \frac{n_1}{\lambda_1} - \sum_{i=1}^{n_1} t_{i:n} + (\alpha - 1) \sum_{i=1}^{n_1} \frac{t_{i:n} e^{-\lambda_1 t_{i:n}}}{1 - e^{-\lambda_1 t_{i:n}}} - (n - n_1)\tau_1$$

$$+ (\alpha - 1) \sum_{i=n_1+1}^{n} \frac{\tau_1 e^{-\lambda_2 \left(t_{i:n} + \frac{\lambda_1}{\lambda_2} \tau_1 - \tau_1 \right)}}{1 - e^{-\lambda_2 \left(t_{i:n} + \frac{\lambda_1}{\lambda_2} \tau_1 - \tau_1 \right)}} = 0, \tag{2.27}$$

$$\dot{l}_{\lambda_2} = \frac{n - n_1}{\lambda_2} - \sum_{i=1}^{n_1} (t_{i:n} - \tau_1) + (\alpha - 1) \sum_{i=n_1+1}^{n} \frac{(t_{i:n} - \tau_1) e^{-\lambda_2 \left(t_{i:n} + \frac{\lambda_1}{\lambda_2} \tau_1 - \tau_1 \right)}}{1 - e^{-\lambda_2 \left(t_{i:n} + \frac{\lambda_1}{\lambda_2} \tau_1 - \tau_1 \right)}} = 0. \tag{2.28}$$

Clearly, the three Eqs. (2.26)–(2.28) are nonlinear equations, and it is not possible to obtain the explicit solutions in this case. The Newton-Raphson method can be used to obtain the solutions of normal equations, as has been described in case of the Weibull distribution. For any iterative process one needs to start from a good set of initial values. The following method can be used to obtain the initial guesses of the Newton-Raphson algorithm. Note that for a given λ_1 and λ_2, the MLE of α can be obtained as a solution of the Eq. (2.26). If we denote this as $\hat{\alpha}(\lambda_1, \lambda_2)$, then

$$\hat{\alpha}(\lambda_1, \lambda_2) = -\frac{n}{\sum_{i=1}^{n_1} \ln(1 - e^{-\lambda_1 t_{i:n}}) + \ln\left(1 - e^{-\lambda_2(t_{i:n} + \frac{\lambda_1}{\lambda_2}\tau_1 - \tau_1)}\right)}. \tag{2.29}$$

Hence, we can obtain the profile log-likelihood function of λ_1 and λ_2 from Eq. (2.25) as $l(\hat{\alpha}(\lambda_1, \lambda_2), \lambda_1, \lambda_2)$. Therefore, from the contour plot of the profile log-likelihood function one can obtain the initial guesses of λ_1 and λ_2, which along with Eq. (2.29) can be used to obtain initial values for α. Since the MLEs cannot be obtained in explicit forms, construction of exact confidence intervals is not possible. However, asymptotic distribution of the MLEs can easily be obtained in terms of the observed Fisher information matrix and that can be used to construct the asymptotic confidence intervals of α, λ_1, and λ_2. Along the same lines as the Weibull distribution, the percentile bootstrap confidence intervals can also be constructed.

2.5.2 Bayesian inference

Samanta and Kundu [78] provided the Bayesian inference of the preceding step-stress model under a fairly flexible set of priors of the unknown parameters. In this case it is assumed that α has a gamma prior with the shape parameter $a_0 > 0$ and the scale parameter $b_0 > 0$, with the following PDF:

$$\pi(\alpha | a_0, b_0) = \begin{cases} 0 & \text{if } \alpha < 0 \\ \frac{b_0^{a_0}}{\Gamma(a_0)} \alpha^{a_0 - 1} e^{-\alpha b_0} & \text{if } \alpha \geq 0. \end{cases} \tag{2.30}$$

To provide dependence between the priors of λ_1 and λ_2, the following transformation has been taken: $\lambda_1 = \beta \lambda_2$. Now, let us consider the new set of parameters as $(\alpha, \theta_2, \beta)$, where $\beta > 0$. It is assumed that θ_2 also has a gamma prior with the shape parameter $a_1 > 0$ and the scale parameter $b_1 > 0$. It is denoted by $\pi(\lambda_2 | a_1, b_1)$, and the PDF has the same form as in Eq. (2.30). Further, it is assumed that β has a Type-II-beta prior with the parameters $a_2 > 0$ and $b_2 > 0$ and it has the following PDF:

$$\pi(\beta | a_2, b_2) = \begin{cases} 0 & \text{if } \beta < 0 \\ \frac{\Gamma(a_2 + b_2)}{\Gamma(a_2)\Gamma(b_2)} \beta^{a_2 - 1} (1 + \beta)^{-(a_2 + b_2)} & \text{if } \beta \geq 0. \end{cases} \tag{2.31}$$

The priors of α, λ_2, and β are assumed to be independently distributed. It may be mentioned that the Type-II-beta distribution with the PDF (2.31) can take a variety of shapes depending on a_2 and b_2. See for example the monograph on beta distributions by Gupta and Nadarajah [79].

The joint prior distribution of λ_1 and λ_2 becomes

$$\pi(\lambda_1, \lambda_2 | a_1, b_1, a_2, b_2) = \begin{cases} ce^{-b_1 \lambda_2} \lambda_1^{a_2 - 1} \lambda_2^{a_1 + b_2} (\lambda_1 + \lambda_2)^{-(a_2 + b_2)} & \text{if } \lambda_1 > 0, \lambda > 0 \\ 0 & \text{otherwise,} \end{cases} \tag{2.32}$$

where $c = \frac{\Gamma(a_2+b_2)}{\Gamma(a_2)\Gamma(b_2)} \times \frac{b_1^{a_1}}{\Gamma(a_1)}$ is the normalizing constant. It is immediate from Eq. (2.32) that λ_1 and λ_2 are dependent for all values of a_1, b_1, a_2, and b_2. Now we provide the BE and the associated HPD CRI of $g(\alpha, \lambda_2, \beta)$, a function of the unknown parameters, under the squared error loss function. In this case also, although we are considering the squared error loss function, any other loss function can also be easily incorporated.

Based on the complete data (Eq. 2.24), and using the priors on α, λ_2, and β as described previously, the joint posterior PDF for $\alpha > 0$, $\lambda_2 > 0$, and $\beta > 0$ can be written as

$$\pi(\alpha, \lambda_2, \beta | \mathcal{D}) \propto C(\beta) \lambda_2^{n+b_1-1} e^{-A_1(\beta)\lambda_2} \alpha^{n+b_0-1} e^{-A_2(\beta,\lambda_2)\alpha} \prod_{i=1}^{n_1} \left(1 - e^{-\beta \lambda_2 t_{i:n}}\right)^{-1}$$

$$\times \prod_{i=n_1+1}^{n} \left(1 - e^{-\lambda_2(t_{i:n}-\tau_1+\tau_1\beta)}\right)^{-1},$$

where

$$C(\beta) = \beta^{n_1+a_2-1}(1+\beta)^{a_2+b_2},$$

$$A_1(\beta) = a_1 + \beta \sum_{i=1}^{n_1} t_{i:n} + \sum_{i=n_1+1}^{n} (t_{i:n} - \tau_1 + \tau_1\beta),$$

and

$$A_2(\beta, \lambda_2) = a_0 - \sum_{i=1}^{n_1} \ln\left(1 - e^{-\beta \lambda_2 t_{i:n}}\right) - \sum_{i=n_1+1}^{n} \left(1 - e^{-\lambda_2(t_{i:n}-\tau_1+\tau_1\beta)}\right).$$

Hence, $\hat{g}_{BE}(\alpha, \lambda_2, \beta)$, the BE of $g(\alpha, \lambda_2, \beta)$, can be obtained as the posterior expectation of $g(\alpha, \lambda_2, \beta)$, as follows:

$$\hat{g}_{BE}(\alpha, \lambda_2, \beta) = \frac{\int_0^1 \int_0^\infty \int_0^\infty g(\alpha, \lambda_2, \beta)\pi(\alpha, \lambda_2, \beta | \mathcal{D}) d\beta d\lambda_2 d\alpha}{\int_0^1 \int_0^\infty \int_0^\infty \pi(\alpha, \lambda_2, \beta | \mathcal{D}) d\beta d\lambda_2 d\alpha}, \tag{2.33}$$

if the expectation exists. Clearly, Eq. (2.33) cannot be obtained in explicit form for a general $g(\alpha, \lambda_2, \beta)$. It is possible to use Lindley's [67] approximation to approximate Eq. (2.33), but it is not possible to compute the associated HPD CRI of $g(\alpha, \lambda_2, \beta)$ using that approximation. Hence, it is not pursued here. Alternatively, it is possible to use a very effective importance sampling technique to compute a simulation consistent estimate of Eq. (2.33), and also to construct the associated HPD CRI.

First observe that the posterior PDF of α, λ_2, and β can be written as follows:

$$\pi(\alpha, \lambda_2, \beta | \mathcal{D} \propto h(\alpha, \lambda_2, \beta | \mathcal{D})\pi_1(\beta | \mathcal{D})\pi_2(\lambda_2 | \beta, \mathcal{D})\pi_3(\alpha | \lambda_2, \beta, \mathcal{D}),$$

where

$$h(\alpha, \lambda_2, \beta | \mathcal{D}) = \beta^{n_1} [A_1(\beta)]^{-(n+b_1)} [A_2(\beta, \lambda_2)]^{-(n+b_0)} \prod_{i=1}^{n_1} \left(1 - e^{-\beta \lambda_2 t_{i:n}}\right)^{-1}$$

$$\times \prod_{i=n_1+1}^{n} \left(1 - e^{-\lambda_2(t_{i:n} - \tau_1 + \tau_1 \beta)}\right)^{-1},$$

$$\pi_1(\beta | \mathcal{D}) = \frac{\Gamma(a_2 + b_2 + n_1)}{\Gamma(a_2 + n_1)\Gamma(b_2)} \beta^{a_2 + n_1 - 1} (1 + \beta)^{-(a_2 + b_2 + n_1)},$$

$$\pi_2(\lambda_2 | \beta, \mathcal{D}) = \frac{[A_1(\beta)]^{n+b_1}}{\Gamma(n + b_1)} \lambda_2^{n+b_1 - 1} e^{-A_1(\beta)\lambda_2},$$

$$\pi_3(\alpha | \lambda_2, \beta, \mathcal{D}) = \frac{[A_2(\beta, \lambda_2)]^{n+b_0}}{\Gamma(n + b_0)} \alpha^{n+b_0 - 1} e^{-A_2(\beta, \lambda_2)\alpha}.$$

Now, the following algorithm can be used to compute a simulation consistent estimate of Eq. (2.33), and the associated HPD CRI.

Algorithm 2.3

1. (a) Generate β_1 from a Type-II-beta distribution with the parameters $a_2 + n_1$ and $b_2 + n_1$, (b) generate λ_{21} from a gamma distribution with the shape parameter $n + b_1$ and the scale parameter $A_1(\beta_1)$, (c) and generate α_1 from a gamma distribution with the shape and scale parameters $n + b_0$ and $A_2(\beta_1, \lambda_{21})$, respectively.
2. Repeat Step 1 N times to obtain $\{(\beta_1, \lambda_{21}, \alpha_1), \ldots, (\beta_N, \lambda_{2N}, \alpha_N)\}$.
3. Calculate $g_i = g(\alpha_i, \lambda_{2i}, \beta_i)$, and $w_i = \dfrac{h(\alpha_i, \lambda_{2i}, \beta_i)}{\sum_{k=1}^{N} h(\alpha_k, \lambda_{2k}, \beta_k)}$.
4. An estimate of Eq. (2.33) can be obtained as $\sum_{i=1}^{N} w_i g_i$.
5. Rearrange $(g_1, w_1), \ldots, (g_N, w_N)$, as $(g_{(1)}, w_{(1)}), \ldots, (g_{(N)}, w_{(N)})$, where $g_{(1)} \leq g_{(2)} \leq \cdots \leq g_{(N)}$. Note that $w_{(i)}$'s are not ordered, they are just associated with $g_{(i)}$'s.
6. A $100(1 - \gamma)\%$ CRI of $g(\alpha, \lambda_2, \beta)$ can be obtained as (g_{j_1}, g_{j_2}), where j_1, j_2 satisfy the following conditions:

$$j_1, j_2 \in \{1, 2, \ldots, N\}, \quad j_1 < j_2, \quad \sum_{i=j_1}^{j_2} w_{(i)} \leq 1 - \gamma < \sum_{i=j_1}^{j_2+1} w_{(i)}. \tag{2.34}$$

7. The $100(1 - \gamma)\%$ HPD CRI of $g(\alpha, \lambda_2, \beta)$ can be obtained as $(g_{j_1^*}, g_{j_2^*})$, where $1 \leq j_1^* < j_2^* \leq N$ satisfy the following conditions:

$$\sum_{i=j_1^*}^{j_2^*} w_{(i)} \leq 1 - \gamma < \sum_{i=j_1^*}^{j_2^*+1} w_{(i)}, \quad \text{and} \quad g_{j_2^*} - g_{j_1^*} \leq g_{j_2} - g_{j_1},$$

for all j_1 and j_2, satisfying Eq. (2.34).

2.5.3 Simulation experiments

Samanta and Kundu [78] performed extensive simulation experiments for different sample sizes, different parameters and τ_1 values to compare the performances of the MLEs and the BEs in terms of their biases and the MSEs. Another aim of the experiments is to compare the performances of the different confidence and CRIs in terms of their coverage percentages and average lengths. It should be mentioned that the BEs are obtained based on almost noninformative priors as suggested by Congdon [80]. In this case it is assumed that $a_0 = b_0 = a_1 = b_1 = 0.0001$, and $a_2 = b_2 = 1$. We briefly report their findings in the following.

It is observed that the biases and the MSEs of the MLEs are significantly larger than those of the BEs. Moreover, the average lengths of the confidence intervals are larger than the average lengths of the corresponding HPD CRIs. The coverage percentages of the HPD CRIs are quite close to the nominal value. Hence, BEs with noninformative priors are recommended in this case, at least for small or moderate sample sizes. For large sample sizes, it does not make any difference.

2.5.4 Data analysis

We present the analysis of a simple step-stress data set. A simple step-stress experiment has been conducted in order to assess the reliability characteristics of a solar lighting device. In this case the temperature is the stress factor, and its level was changed during the experiment from the normal operating temperature 293K (Kelvin) to 353K. Thirty one (31) devices are put on a life test at the initial temperature 293K, and then the temperature is changed to 353K at the time point $\tau_1 = 5$ (in hundred hours). The data are presented in Table 2.5.

In this case the experiment continues till all the device fail, and before time τ_1, the number of failures is $n_1 = 16$. The data set has been analyzed by Samanta and Kundu [78] based on the assumption that the lifetime distributions of the solar lighting device follow generalized exponential distributions with the same shape parameter at the two different stress levels, but with different scale parameter. The contour plot of the profile likelihood function of λ_1 and λ_2 is provided in Fig. 2.3. It is clear from the contour that the initial guesses of λ_1 and λ_2 can be taken as 0.20 and 3.50, respectively. From these initial guesses, using Eq. (2.29) the initial guess of α can be obtained as 3.523.

Table 2.5 **Solar lighting device data set of size 31**

Stress level	Data								
S_1	0.140	0.783	1.324	1.582	1.716	1.794	1.883	2.293	2.660
	2.674	2.725	3.085	3.924	4.396	4.612	4.892		
S_2	5.002	5.022	5.082	5.112	5.147	5.238	5.244	5.247	5.305
	5.337	5.407	5.408	5.445	5.483	5.717			

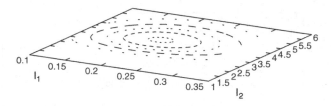

Fig. 2.3 The contour plot of the profile likelihood function of λ_1 and λ_2.

Table 2.6 **Asymptotic confidence intervals of α, λ_1, and λ_2 for the solar lighting device data set**

	Unknown parameters					
	α		θ_1		θ_2	
Level	LL	UL	LL	UL	LL	UL
90%	0.7244	2.2464	0.0894	0.3176	2.2160	5.3388
95%	0.5760	2.3949	0.0671	0.3398	1.9113	5.6434
99%	0.2929	2.6779	0.0247	0.3823	1.3306	6.2242

Based on these initial guesses using the Newton-Raphson algorithm the MLEs of α, θ_1, and θ_2 are obtained as 1.4854, 0.2035, and 3.7774, respectively. In Table 2.6, 90%, 95%, and 99% asymptotic confidence intervals of the unknown parameters are presented. The BEs of α, θ_1, and θ_2 are, respectively, 1.4577, 0.1999, and 3.4180. Different symmetric and HPD CRIs are presented in Table 2.7.

One natural question arises whether generalized exponential distribution provides a good fit to the data set or not. The empirical survival function and the fitted survival functions based on MLEs and the BEs are provided in Fig. 2.4. The fitted survival functions are almost indistinguishable at least before τ_1. The Kolmogorov-Smirnov (KS) distance between the observed and the fitted distribution functions are obtained. It is observed that the KS distance between the empirical CDF and the fitted CDF based on MLEs is 0.1121 and the associated p-value is 0.8311. In case of the fitted

Table 2.7 **Symmetric and HPD CRIs of α, λ_1, and λ_2 for the solar lighting device data set**

		Unknown parameters					
		α		θ_1		θ_2	
CRI	Level	LL	UL	LL	UL	LL	UL
	90%	0.8618	2.3402	0.1068	0.3171	2.1320	5.0691
Symm.	95%	0.7923	2.4172	0.0904	0.3189	1.9169	5.4008
	99%	0.6629	2.6146	0.0668	0.3430	1.5888	5.7524
	90%	0.7923	2.1200	0.0946	0.2888	2.0641	4.8812
HPD	95%	0.7624	2.3714	0.0950	0.3195	1.6664	5.0951
	99%	0.6629	2.5845	0.0681	0.3430	1.6149	5.7524

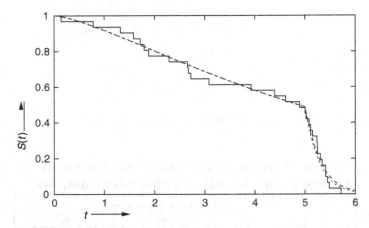

Fig. 2.4 Empirical survival function and fitted survival functions based on MLEs and BEs of the solar lighting device data set.

BEs, the KS distance is 0.0939 and the corresponding p value is 0.9475. Therefore, it is clear that the generalized exponential distribution provides a good fit to the data set and between the two estimators, BEs provide a slightly better fit in terms of the lower KS distance.

2.5.5 Open problems

In this section we indicate some open problems related to the generalized exponential distribution step-stress model for further consideration.

Open Problem: Most of the problems which have been mentioned for the Weibull distribution in Section 2.4.6 are possible open problems for the generalized exponential distribution also.

Open Problem: Let us recall that the two-parameter generalized exponential distribution is a special case of the three-parameter exponentiated Weibull distribution, introduced by Mudholkar and Srivastava [71]. The three-parameter exponentiated Weibull distribution is more flexible than the two-parameter generalized exponential distribution due to the presence of two shape parameters. It is worth investigating the step-stress model when the lifetime of the experimental units follow the three-parameter exponentiated Weibull distribution.

Open Problem: Prior elicitation is an important problem in general. Although Samanta and Kundu [78] developed the Bayesian inference under a fairly general set of priors on the unknown parameters, no work has been done in choosing the proper priors. More work is needed in that direction.

Open Problem: It is well known that discriminating between a Weibull and a generalized exponential distribution is a difficult problem. Gupta and Kundu [81] provided a detailed analysis of the probability of correct selection in discriminating between these two distributions for complete sample. It will be important to develop the necessary methodology for step-stress data.

2.6 Other continuous distributions

In this section we provide briefly the inferential procedures for some other continuous lifetime distributions, viz., (i) gamma distribution, (ii) log-normal distribution, (iii) log-logistic distribution, (iv) Pareto distribution, and (v) Birnbaum-Saunders distribution.

2.6.1 Gamma distribution

The gamma distribution has been used quite extensively in reliability and survival analysis for analyzing lifetime data. In this section it is assumed that a two-parameter gamma distribution with the shape parameter $\beta > 0$ and the scale parameter $\theta > 0$ has the PDF of the following form:

$$f(t;\alpha,\theta) = \begin{cases} \frac{1}{\Gamma(\alpha)\theta^\alpha} t^{\alpha-1} e^{-\frac{t}{\theta}} & \text{if } t > 0 \\ 0 & \text{if } t \le 0. \end{cases}$$

The corresponding CDF can be written as

$$F(t;\alpha,\theta) = \begin{cases} IG_\alpha\left(\frac{t}{\theta}\right) & \text{if } t > 0 \\ 0 & \text{if } t \le 0, \end{cases}$$

where $IG_\alpha(t)$ is the incomplete gamma ratio as given here:

$$IG_\alpha(t) = \int_0^t \frac{1}{\Gamma(\alpha)} x^{\alpha-1} e^{-x} dx, \quad t > 0, \quad \alpha > 0.$$

The two-parameter gamma distribution is an extension of a one-parameter exponential distribution and is a very flexible distribution, like a two-parameter Weibull distribution. The PDF of a two-parameter gamma distribution can be a decreasing or a unimodal function provided the shape parameter is less than one or greater than one, respectively. The hazard function of a two-parameter gamma distribution can be an increasing or a decreasing function depending on the shape parameter. If the shape parameter is greater than one, the hazard function is an increasing function, otherwise it is a decreasing function. Many properties of a two-parameter gamma distribution are similar to the corresponding properties of a two-parameter Weibull distribution. In fact if the sample size is small, it is very difficult to judge whether the data are coming from a gamma or from a Weibull distribution; see for example Fearn and Nebenzahl [82]. One major advantage of a gamma distribution is that it belongs to a two-dimensional exponential family. For detailed discussions and for several important properties of a gamma distribution, see the book by Bowman and Shenton [83].

Alkhalfan [33] in her PhD thesis considered the classical inference of the simple step-stress model when the lifetime distribution of the experimental units follow gamma distribution with the common shape parameter at the two different stress levels but different scale parameters. She has considered Type-I, Type-II, and Type-II progressive censoring schemes. Moreover, she has extended most of the results for the multiple step-stress model. We will be presenting the case in detail when the data are Type-II censored; for other cases the readers are referred to the original thesis of the author.

In this section we make the following assumptions and use these notations. It is assumed that the lifetime distribution of the experimental units follow a gamma distribution with the scale parameter $\theta_j > 0$ at the stress level s_j, for $j = 1$ and 2, and the shape parameter $\alpha > 0$ at both the stress levels. Hence, based on the CEM assumption, the PDF of an experimental unit under a simple SSLT can be written as

$$f(t) = \begin{cases} g_1(t) & \text{if } 0 \le t < \tau_1 \\ g_2(t) & \text{if } \tau_1 \le t < \infty, \end{cases} \tag{2.35}$$

where

$$g_1(t) = \frac{1}{\theta_1^\alpha} t^{\alpha-1} e^{-\frac{t}{\theta_1}},$$

$$g_2(t) = \frac{1}{\theta_2^\alpha} \left(t - \tau_1 + \frac{\theta_2}{\theta_1} \tau_1 \right)^{\alpha-1} e^{-\frac{1}{\theta_2}\left(t - \tau_1 + \frac{\theta_2}{\theta_1}\tau_1\right)}.$$

The corresponding CDF becomes

$$F(t) = \begin{cases} IG_\alpha\left(\frac{t}{\theta_1}\right) & \text{if } 0 \le t < \tau_1 \\ IG_\alpha\left(\frac{t - \tau_1 + \frac{\theta_2}{\theta_1}\tau_1}{\theta_2}\right) & \text{if } \tau_1 \le t < \infty. \end{cases}$$

Hence, the hazard function of $F(t)$ is given by

$$h(t) = \begin{cases} \dfrac{g_1(t)}{1 - IG_\alpha\left(\frac{t}{\theta_1}\right)} & \text{if } 0 < t < \tau_1 \\ \dfrac{g_2(t)}{1 - IG_\alpha\left(\frac{t - \tau_1 + \frac{\theta_2}{\theta_1}\tau_1}{\theta_2}\right)} & \text{if } \tau_1 \le t < \infty. \end{cases}$$

Here $g_1(t)$ and $g_2(t)$ are the same as defined in Eq. (2.35). In this case also when $\alpha = 1$,

$$h(t) = \begin{cases} \frac{1}{\theta_1} & \text{if } 0 < t < \tau_1 \\ \frac{1}{\theta_2} & \text{if } \tau_1 \le t < \infty, \end{cases}$$

since $IG_1(t) = 1 - e^{-t}$. Now we provide the MLEs of the unknown parameters α, λ_1, and λ_2, based on the Type-II censored data of the form Eq. (2.18). The MLEs of α, θ_1, and θ_2 exist only when $1 \le n_1 \le r - 1$. Hence, we discuss the conditional MLEs of α, θ_1, and θ_2, conditioning on the fact that $1 \le N_1 \le r - 1$. Based on the CEM assumption, the log-likelihood function without the additive constant can be written as

$$l(\alpha, \theta_1, \theta_2) = -r \ln \Gamma(\alpha) - \alpha n_1 \Gamma(\alpha) - \alpha n_1 \ln \theta_1 - \alpha(r - n_1) \ln \theta_2 - \sum_{i=1}^{n_1} \frac{t_{i:n}}{\theta_1}$$

$$- \frac{1}{\theta_2} \sum_{i=n_1+1}^{r} \left(t_{i:n} - \tau_1 + \frac{\theta_2}{\theta_1} \tau_1 \right) + (\alpha - 1) \sum_{i=n_1+1}^{r} \ln \left(t_{i:n} - \tau_1 + \frac{\theta_2}{\theta_1} \tau_1 \right)$$

$$+ (n - r) \ln \left(1 - IG_\alpha \left(t_{r:n} - \tau_1 + \frac{\theta_2}{\theta_1} \tau_1 \right) \right). \tag{2.36}$$

The MLEs can be obtained by maximizing Eq. (2.36) with respect to α, θ_1, and θ_2. The corresponding normal equations can be obtained by differentiating $l(\alpha, \theta_1, \theta_2)$ with respect to α, θ_1, and θ_2 respectively, and they are provided as follows:

$$\dot{l}_\alpha = -r\Psi(\alpha) - n_1 \ln \theta_1 - (r - n_1) \ln \theta_2 + \sum_{i=1}^{n_1} \ln t_{i:n} + \sum_{i=n_1+1}^{r} \ln \left(t_{i:n} - \tau_1 + \frac{\theta_2}{\theta_1} \tau_1 \right)$$

$$+ \frac{n - r}{1 - IG_\alpha(s)} \left[\Psi(\alpha)IG_\alpha(s) - \frac{1}{\Gamma(\alpha)} \int_0^s u^{\alpha-1} \ln(u)e^{-u} du \right] = 0, \tag{2.37}$$

$$\dot{l}_{\theta_1} = -\frac{\alpha n_1}{\theta_1} + \frac{(r - n_1)\tau_1}{\theta_1^2} - \frac{(\alpha - 1)\theta_2 \tau_1}{\theta_1^2} \sum_{i=n_1+1}^{r} \frac{1}{t_{i:n} - \tau_1 + \frac{\theta_2 \tau_1}{\theta_1}}$$

$$+ \frac{1}{\theta_1^2} \sum_{i=1}^{n_1} t_{i:n} + \frac{(n - r)\tau_1 s^{\alpha-1} e^{-s}}{(1 - IG_\alpha(s))\Gamma(\alpha)\theta_1^2} = 0, \tag{2.38}$$

$$\dot{l}_{\theta_2} = -\frac{\alpha(r - n_1)}{\theta_2} + \frac{1}{\theta_2^2} \sum_{i=n_1+1}^{r} (t_{i:n} - \tau_1) + \frac{(\alpha - 1)\tau_1}{\theta_1} \sum_{i=n_1+1}^{r} \frac{1}{t_{i:n} - \tau_1 + \frac{\theta_2 \tau_1}{\theta_1}}$$

$$+ \frac{(n - r)(t_{r:n} - \tau_1)\tau_1 s^{\alpha-1} e^{-s}}{(1 - IG_\alpha(s))\Gamma(\alpha)\theta_2^2} = 0. \tag{2.39}$$

Here, $\Psi(\alpha)$ denotes the digamma function, i.e., $\Psi(\alpha) = \frac{d}{d\alpha} \ln(\Gamma(\alpha)) = \frac{\Gamma'(\alpha)}{\Gamma(\alpha)}$, and $s = \frac{t_{r:n} - \tau_1}{\theta_2} + \frac{\tau_1}{\theta_1}$. It may be mentioned that there are several well-known approximations available for the digamma function; see for example Srivastava and Choi [84].

Clearly, the normal equations (2.37)–(2.39) do not have any obvious simplifications and they do not have any explicit solutions. One needs to use a numerical technique like the Newton-Raphson or Gauss-Newton method to find the solutions of these nonlinear equations. Alternatively, a nonlinear optimization algorithm may be used to compute the MLEs of the unknown parameters. Alkhalfan [33] suggested using the function *optim* in R to compute the MLEs of the unknown parameters for the two-parameter gamma distribution. To find the initial guess of any iterative procedure, similar choices to the Weibull distribution (see Section 2.4.2) may be used.

To develop the Newton-Raphson method or to obtain the asymptotic variance covariance matrix of the MLEs, one needs to compute the observed Fisher information matrix, and that can be obtained by taking the second derivative of the log-likelihood function as given in Eq. (2.36). The expressions of the Fisher information matrix are available in Alkhalfan [33], and they are not provided here.

The bootstrap confidence intervals and the asymptotic confidence intervals can be used in a routine matter to construct confidence intervals of the unknown parameters. Alkhalfan [33] conducted an extensive simulation experiment to compare the performances of the two different confidence intervals and also to see how the MLEs work in practice for different sample sizes and for different values of the parameters. It is observed that the performance of the MLEs is quite satisfactory particularly if the sample size is not too small. Bootstrap confidence intervals perform better than the asymptotic confidence intervals in the sense that the coverage percentages of the bootstrap confidence intervals are closer to the corresponding nominal levels compared to the asymptotic confidence intervals for small or moderate sample sizes. Hence, in this case bootstrap confidence intervals are recommended for constructing confidence intervals of the unknown parameters for small and moderate sample sizes. For large sample sizes, asymptotic confidence intervals can be used. Now we mention some open problems.

Open Problem: Alkhalfan [33] developed a classical inference of the simple step-stress model when the lifetime distribution of the experimental units follow a gamma distribution for different censoring schemes. It will be important to develop a Bayesian inference of the same model and compare their performances with the classical inference. Moreover, we believe that for the Bayesian inference a general methodology for different censoring schemes should be possible to develop.

Open Problem: Alkhalfan [33] developed the method based on the assumption that at the two stress levels the lifetime of the experimental units follows gamma distributions with the same shape parameter and different scale parameters. A more general model would be when the two shape parameters are not assumed to be the same. It will be interesting to develop the inference procedures, both classical and Bayesian, under this general framework, and then test whether the two shape parameters are equal or not.

2.6.2 Log-normal distribution

The log-normal distribution has been used quite extensively in analyzing lifetime data. If X has a normal distribution then e^X has a log-normal distribution. Therefore, a log-normal distribution with the scale parameter $0 < \lambda < \infty$ and the shape parameter $\sigma > 0$ has the following CDF:

$$F(t; \lambda, \sigma) = \begin{cases} 0 & \text{if } t < 0 \\ \Phi\left(\frac{\ln(t) - \ln(\lambda)}{\sigma}\right) & \text{if } t \geq 0. \end{cases}$$

The corresponding PDF and hazard function become

$$
f(t; \lambda, \sigma) = \begin{cases} 0 & \text{if } t < 0 \\ \frac{1}{\sigma t} \phi\left(\frac{\ln(t) - \ln(\lambda)}{\sigma}\right) & \text{if } t \geq 0, \end{cases}
$$

and

$$
h(t; \lambda, \sigma) = \frac{\phi\left(\frac{\ln(t) - \ln(\lambda)}{\sigma}\right)}{\sigma t \Phi\left(\frac{-\ln(t) + \ln(\lambda)}{\sigma}\right)}; \quad t > 0,
$$

respectively. The PDF and the hazard function of a log-normal distribution are always unimodal functions. The PDF of a log-normal distribution is very similar to the PDFs of gamma, Weibull or generalized exponential distributions when the shape parameters of gamma, Weibull and generalized exponential distributions are greater than one. It has been shown by Kundu and Manglick [85, 86] and Kundu et al. [87] that it is very difficult to discriminate between log-normal and gamma, log-normal and Weibull and log-normal and generalized exponential distributions. For different properties of a log-normal distribution and for its various applications, one is referred to Johnson et al. [59].

Alhadeed [88] considered in his PhD thesis the analysis of the log-normal step-stress model, see also Alhadeed and Yang [34], when the complete data are available. Balakrishnan et al. [55] considered the same problem when the data are Type-I censored. It is assumed that the lifetime distribution of the experimental units at the two different stress levels follow log-normal distributions with different scale parameters, λ_1 and λ_2, but the same shape parameter σ. Based on the CEM assumption, the CDF of the lifetime of an experimental unit from a simple step-stress model can be written as

$$
F(t) = \begin{cases} 0 & \text{if } t < 0 \\ \Phi\left(\frac{\ln(t) - \ln(\lambda_1)}{\sigma}\right) & \text{if } 0 \leq t < \tau_1 \\ \Phi\left(\frac{\ln\left(t + \tau_1 \frac{\lambda_2}{\lambda_1} - \tau_1\right) - \ln(\lambda_2)}{\sigma}\right) & \text{if } \tau_1 \leq t < \infty. \end{cases} \tag{2.40}
$$

Hence, the PDF corresponding to Eq. (2.40) becomes

$$
f(t) = \begin{cases} 0 & \text{if } t < 0 \\ \frac{1}{\sigma t} \phi\left(\frac{\ln(t) - \ln(\lambda_1)}{\sigma}\right) & \text{if } 0 \leq t < \tau_1 \\ \frac{1}{\sigma\left(t + \frac{\lambda_2}{\lambda_1}\tau_1 - \tau_1\right)} \phi\left(\frac{\ln\left(t + \tau_1 \frac{\lambda_2}{\lambda_1} - \tau_1\right) - \ln(\lambda_2)}{\sigma}\right) & \text{if } \tau_1 \leq t < \infty. \end{cases} \tag{2.41}
$$

In this case it is more convenient to work with the log-transformation of the data than the original data. Now if a random variable T has the PDF (2.41), then $Y = \ln(T)$ has the PDF

$$f_Y(y) = \begin{cases} 0 & \text{if } t < 0 \\ \frac{1}{\sigma}\phi\left(\frac{y-\mu_1}{\sigma}\right) & \text{if } 0 < t < \ln\tau_1 \\ \frac{e^y}{\sigma(e^y+e^{\mu_2-\mu_1}\tau_1-\tau_1)}\phi\left(\frac{\ln(e^y+\tau_1 e^{\mu_2-\mu_1}-\tau_1)-\mu_2}{\sigma}\right) & \text{if } \ln\tau_1 \le y < \infty. \end{cases}$$

Here $\mu_1 = \ln\lambda_1$ and $\mu_2 = \ln\lambda_2$. Therefore, if we denote the log of the observed lifetimes as $y_{i:n} = \ln(t_{i:n})$ for $i = 1, \ldots, n$, then the log-likelihood function based on the complete observations $\{y_{1:n}, \ldots, y_{n:n}\}$ is

$$l(\mu_1, \mu_2, \sigma) = -\frac{n}{2}\ln(\pi) - n\ln\sigma - \frac{1}{2}\sum_{i=1}^{n_1}\left(\frac{y_{i:n}-\mu_1}{\sigma}\right)^2$$

$$- \sum_{i=n_1+1}^{n}\ln(e^{y_{i:n}} + \tau_1 e^{\mu_2-\mu_1} - \tau_1)$$

$$- \frac{1}{2}\sum_{i=n_1+1}^{n}\left(\frac{\ln(e^{y_{i:n}}+\tau_1 e^{\mu_2-\mu_1}-\tau_1)-\mu_2}{\sigma}\right)^2. \tag{2.42}$$

Here it is assumed that $1 \le n_1 \le n - 1$ and $n \ge 3$; otherwise it is known that the MLEs of σ, μ_1, and μ_2 do not exist. Therefore, the conditional MLEs of the unknown parameters conditioning on $1 \le N_1 \le n - 1$ can be obtained by maximizing Eq. (2.42) with respect to the unknown parameters. In this case the normal equations become

$$\dot{l}_{\mu_1} = \sum_{i=n_1+1}^{n}\frac{\tau_1 e^{\mu_2-\mu_1}}{e^{y_{i:n}}+\tau_1 e^{\mu_2-\mu_1}-\tau_1} + \frac{1}{\sigma^2}\sum_{i=1}^{n_1}(y_{i:n}-\mu_1)$$

$$+ \frac{1}{\sigma^2}\sum_{i=n_1+1}^{n}(\ln(e^{y_{i:n}}+\tau_1 e^{\mu_2-\mu_1}-\tau_1)-\mu_2)\frac{\tau_1 e^{\mu_2-\mu_1}}{e^{y_{i:n}}+\tau_1 e^{\mu_2-\mu_1}} = 0, \quad (2.43)$$

$$\dot{l}_{\mu_2} = -\frac{1}{\sigma^2}\sum_{i=n_1+1}^{n}(\ln(e^{y_{i:n}}+\tau_1 e^{\mu_2-\mu_1}-\tau_1)-\mu_2)\left[\frac{\tau_1 e^{\mu_2-\mu_1}}{e^{y_{i:n}}+\tau_1 e^{\mu_2-\mu_1}}-1\right]$$

$$- \sum_{i=n_1+1}^{n}\frac{\tau_1 e^{\mu_2-\mu_1}}{e^{y_{i:n}}+\tau_1 e^{\mu_2-\mu_1}-\tau_1} = 0, \tag{2.44}$$

$$\dot{l}_\sigma = -\frac{n}{\sigma} + \frac{1}{\sigma^3}\sum_{i=1}^{n_1}(y_{i:n}-\mu_1)^2$$

$$+ \frac{1}{\sigma^3}\sum_{i=n_1+1}^{n}(\ln(e^{y_{i:n}}+\tau_1 e^{\mu_2-\mu_1}-\tau_1)-\mu_2)^2 = 0. \tag{2.45}$$

Clearly, Eqs. (2.43)–(2.45) cannot be solved explicitly. One needs to use the Newton-Raphson type iterative algorithm to solve Eqs. (2.43)–(2.45) numerically. Some initial guesses of the parameters are needed to start the iteration. If μ_1 and μ_2 are known, the MLE of σ^2 can be obtained from Eq. (2.45) as

$$\hat{\sigma}^2(\mu_1,\mu_2) = \frac{1}{n}\left[\sum_{i=1}^{n_1}(y_{i:n}-\mu_1)^2 + \sum_{i=n_1+1}^{n}(\ln(e^{y_{i:n}}+\tau_1 e^{\mu_2-\mu_1}-\tau_1)-\mu_2)^2\right].$$

(2.46)

We can obtain the profile log-likelihood function of μ_1 and μ_2 by using Eq. (2.46) in Eq. (2.42). The profile log-likelihood function of μ_1 and μ_2 without the additive constants can be written as

$$p(\mu_1,\mu_2) = -\frac{n}{2}\ln\left[\sum_{i=1}^{n_1}(y_{i:n}-\mu_1)^2 + \sum_{i=n_1+1}^{n}(\ln(e^{y_{i:n}}+\tau_1 e^{\mu_2-\mu_1}-\tau_1)-\mu_2)^2\right]$$
$$- \sum_{i=n_1+1}^{n}\ln(e^{y_{i:n}}+\tau_1 e^{\mu_2-\mu_1}-\tau_1).$$

(2.47)

A contour plot of $p(\mu_1,\mu_2)$ as in Eq. (2.47) may provide good starting values of μ_1 and μ_2. Once we obtain the starting values of μ_1 and μ_2, the starting value of σ can be easily obtained from Eq. (2.46). Although we have presented the results here for the complete sample, similar results can be developed for different censoring schemes. Balakrishnan et al. [55] performed an extensive simulation study to compare the performances of different confidence intervals. It is observed that the biased corrected bootstrap method works very well in this case. Most of the results have been extended by Lin and Chou [56] for the multiple step-stress model.

2.6.3 Log-logistic distribution

The log-logistic distribution has been used quite frequently to analyze positively skewed data. The two-parameter log-logistic distribution has the following CDF:

$$F(t;\alpha,\lambda) = \begin{cases} 0 & \text{if } t < 0 \\ \dfrac{(t/\lambda)^\alpha}{1+(t/\lambda)^\alpha} & \text{if } 0 \le t < \infty. \end{cases}$$

(2.48)

The corresponding PDF becomes

$$f(t;\alpha,\lambda) = \begin{cases} 0 & \text{if } t < 0 \\ \dfrac{\alpha(t/\lambda)^{\alpha-1}}{\lambda\left(1+(t/\lambda)^\alpha\right)^2} & \text{if } 0 \le t < \infty. \end{cases}$$

Here $\lambda > 0$ and $\alpha > 0$ are the scale and shape parameters, respectively. Moreover, λ is the median of the log-logistic distribution with the CDF (2.48). The shapes of the PDF and the hazard functions depend on the shape parameter α. The PDF and the hazard function of a log-logistic distribution is either a decreasing or a unimodal function. For different properties of a two-parameter log-logistic distribution, interested readers are referred to Johnson et al. [59]. For certain ranges of the parameters, the shape of the PDFs of log-logistic and log-normal distributions can be very similar. Therefore, often it is very difficult to discriminate between a log-normal and a log-logistic distribution if the sample size is not very large; see for example Dey and Kundu [89] in this respect.

Srivastava and Shukla [90] considered the simple step-stress model when the lifetime of an experimental unit follows log-logistic distribution with the same shape parameter α but having different scale parameters λ_1 and λ_2, at the two different stress levels; see also Al-Masri and Ebrahem [91] and Ebrahem and Al-Masri [92].

The CDF of the lifetime of an experimental unit from this step-stress model based on the CEM assumption is

$$F(t) = \begin{cases} 0 & \text{if } t < 0 \\ F_1(t) & \text{if } 0 \le t < \tau_1 \\ F_2(t - \tau_1 + (\lambda_2/\lambda_1)\tau_1) & \text{if } \tau_1 \le t < \infty, \end{cases}$$

where for $i = 1, 2$ $F_i(t) = \frac{(t/\lambda_i)^\alpha}{1+(t/\lambda_i)^\alpha}$ for $t > 0$. The PDF of a test unit from the previous simple step-stress model becomes

$$f(t) = \begin{cases} 0 & \text{if } t < 0 \\ g_1(t) & \text{if } 0 \le t < \tau_1 \\ g_2(t - \tau_1 + (\lambda_2/\lambda_1)\tau_1) & \text{if } \tau_1 \le t < \infty, \end{cases}$$

where for $t > 0$, and for $i = 1, 2$,

$$g_i(t) = \frac{(\alpha/\lambda_i)(t/\lambda_i)^{\alpha-1}}{(1 + (t/\lambda_i)^\alpha)^2}.$$

Based on the complete observations the problem is to estimate the unknown parameters α, λ_1, and λ_2. It is often assumed that the median life of a test unit is a log-linear function of the stress, i.e.,

$$\ln \lambda_i = \beta_0 + \beta_1 s_i; \quad i = 1, 2.$$

Therefore, the problem now is to estimate the unknown parameters α, β_0, and β_1. Based on the complete observation $\{t_{1:n} < t_{2:n} < \cdots < t_{n_1:n} < \tau_1 < t_{n_1+1:n} < \cdots < t_{n:n}\}$, the log-likelihood function can be written as

$$l(\alpha, \beta_0, \beta_1) = n \ln \alpha - n_1(\beta_0 + \beta_1 s_1) - 2 \sum_{i=1}^{n_1} \ln \left(1 + t_{i:n}^{\alpha} e^{-\alpha(\beta_0 + \beta_1 s_1)}\right)$$

$$+ (\alpha - 1) \left[\sum_{i=1}^{n} \ln t_{i:n} - n_1(\beta_0 + \beta_1 s_1) - (n - n_1)(\beta_0 - \beta_1 s_2)\right]$$

$$- 2 \sum_{i=n_1+1}^{n} \ln \left(1 + t_{i:n}^{\alpha} - \tau_1^{\alpha}\right) e^{-\alpha(\beta_0 + \beta_1 s_2)} + (n - n_1)\tau_1^{\alpha} e^{-\alpha(\beta_0 + \beta_1 s_1)}.$$

$$(2.49)$$

The MLEs of the unknown parameters can be obtained by maximizing Eq. (2.49) with respect to the unknown parameters. The normal equations can be obtained in a routine manner by differentiating the log-likelihood function (2.49) with respect to the unknown parameters and equating them to zero. The MLEs cannot be obtained in closed forms, and some iterative procedure like the Newton-Raphson method may be used to solve these nonlinear equations to compute the MLEs. The asymptotic distribution of the MLEs or the bootstrap method can be used to construct confidence intervals of the unknown parameters.

2.6.4 Pareto distribution

The Pareto distribution was originally introduced by Pareto [93] to model income distribution of a society. The two-parameter Pareto distribution has the following PDF:

$$f(t; \alpha, \theta) = \begin{cases} 0 & \text{if } t < 0 \\ \frac{\alpha \theta^{\alpha}}{(\theta + t)^{\alpha+1}} & \text{if } 0 \leq t < \infty. \end{cases}$$

Here $\alpha > 0$ and $\theta > 0$ are the shape and scale parameters, respectively. The corresponding CDF becomes

$$F(t; \alpha, \theta) = \begin{cases} 0 & \text{if } t < 0 \\ 1 - \frac{\theta^{\alpha}}{(\theta + t)^{\alpha}} & \text{if } 0 \leq t < \infty, \end{cases}$$

and for $t > 0$, the hazard function takes the following form:

$$h(t; \alpha, \theta) = \frac{\alpha}{(\theta + t)}.$$

For all values of α and θ, the PDF and hazard function are decreasing functions. The Pareto distribution is a heavy tail distribution and the mean and variance are not finite for all $\alpha > 0$. Although, it has been extensively used for analyzing income data, due to its simple form it has also been used quite effectively in analyzing heavy tail lifetime data. For different properties of a Pareto distribution and for its different applications, interested readers are referred to Arnold [94].

Kamal et al. [36] considered the simple step-stress model when the lifetime distributions at the two different stress levels follow Pareto distributions with the same shape parameter α but different scale parameters θ_1 and θ_2, respectively. Based on the CEM assumption, the CDF of an experimental unit from this step-stress experiment takes the form:

$$
F(t) = \begin{cases}
0 & \text{if } t < 0 \\
1 - \dfrac{\theta_1^\alpha}{(\theta_1 + t)^\alpha} & \text{if } 0 \le t < \tau_1 \\
1 - \dfrac{\theta_2^\alpha}{\left[\theta_2 + \tau_1 \left(\dfrac{\theta_2}{\theta_1} - 1 \right) + t \right]^\alpha} & \text{if } \tau_1 \le t < \infty.
\end{cases}
$$

The associated PDF and the hazard function are

$$
f(t) = \begin{cases}
0 & \text{if } t < 0 \\
\dfrac{\alpha \theta_1^\alpha}{(\theta_1 + t)^{\alpha+1}} & \text{if } 0 \le t < \tau_1 \\
\dfrac{\alpha \theta_2^\alpha}{\left[\theta_2 + \tau_1 \left(\dfrac{\theta_2}{\theta_1} - 1 \right) + t \right]^{\alpha+1}} & \text{if } \tau_1 \le t < \infty,
\end{cases}
$$

and

$$
h(t) = \begin{cases}
\dfrac{\alpha}{\theta_1 + t} & \text{if } 0 < t < \tau_1 \\
\dfrac{\alpha}{\theta_2 + \tau_1 \left(\dfrac{\theta_2}{\theta_1} - 1 \right) + t} & \text{if } \tau_1 \le t < \infty,
\end{cases}
$$

respectively. Based on the complete observations and using the same notations as before, the log-likelihood function of α, θ_1, and θ_2, can be written as

$$
l(\alpha, \theta_1, \theta_2) = n \ln \alpha + n_1 \alpha \ln \theta_1 + (n - n_1) \alpha \ln \theta_2 - (\alpha + 1) \sum_{i=1}^{n_1} \ln(\theta_1 + t_{i:n})
$$

$$
- (\alpha + 1) \sum_{i=n_1+1}^{n} \ln \left[\theta_2 + \tau_1 \left(\frac{\theta_2}{\theta_1} - 1 \right) + t_{i:n} \right]. \tag{2.50}
$$

The MLEs of the unknown parameters can be obtained by maximizing Eq. (2.50) with respect to the unknown parameters. The normal equations can be obtained in a routine manner and they can be solved using the Newton-Raphson method. Note that when θ_1 and θ_2 are known, the MLE of α, say $\hat{\alpha}(\theta_1, \theta_2)$, can be obtained from the normal equation $\dot{l}_\alpha = 0$ in explicit form as

$$
\hat{\alpha}(\theta_1, \theta_2) = \frac{n}{g(\theta_1, \theta_2)},
$$

where

$$g(\theta_1, \theta_2) = n_1 \ln \theta_1 + (n - n_1) \ln \theta_2 - \sum_{i=1}^{n_1} \ln(\theta_1 + t_{i:n})$$

$$- \sum_{i=n_1+1}^{n} \ln\left[\theta_2 + \tau_1 \left(\frac{\theta_2}{\theta_1} - 1\right) + t_{i:n}\right].$$

Therefore, the MLEs of θ_1 and θ_2 can be obtained by maximizing the profile log-likelihood function of θ_1 and θ_2

$$p(\theta_1, \theta_2) = l(\hat{\alpha}(\theta_1, \theta_2), \theta_1, \theta_2), \tag{2.51}$$

directly by using some optimization package. Alternatively, the contours of Eq. (2.51) can provide very good starting values of the unknown parameters in the Newton-Raphson method. Kamal et al. [36] used the observed Fisher information matrix to construct confidence intervals of the unknown parameters. Simulation results suggest the performances are quite satisfactory in the sense the coverage percentages of the proposed confidence intervals are quite close to the corresponding nominal values for moderate or large sample sizes.

The following problem can be considered as an important open problem related to the Pareto step-stress model.

Open Problem: Note that in this case the mean and variance of a Pareto distribution do not exist for all $\alpha > 0$. It may not satisfy all the regularity conditions which are needed for the MLEs to follow an asymptotically normal distribution. Therefore, it is important to derive the asymptotic distribution of the MLEs in a formal manner.

2.6.5 Birnbaum-Saunders distribution

Birnbaum and Saunders [95, 96] introduced the Birnbaum-Saunders distribution as a failure time distribution for fatigue failure caused due to cyclic loading. A more general derivation was provided by Desmond [97] based on a biological model. A two-parameter Birnbaum-Saunders distribution has the following CDF:

$$F(t; \alpha, \beta) = \begin{cases} 0 & \text{if } t < 0 \\ \Phi\left[\frac{1}{\alpha}\left\{\left(\frac{t}{\beta}\right)^{1/2} - \left(\frac{\beta}{t}\right)^{1/2}\right\}\right] & \text{if } 0 \le t < \infty. \end{cases}$$

Here $\alpha > 0$ and $\beta > 0$ are the shape and scale parameters, respectively. The PDF of a Birnbaum-Saunders distribution is

$$f(t; \alpha, \beta) = \begin{cases} 0 & \text{if } t < 0 \\ \frac{1}{2\sqrt{2\pi}\alpha\beta}\left[\left(\frac{\beta}{t}\right)^{1/2} + \left(\frac{\beta}{t}\right)^{3/2}\right] \exp\left[-\frac{1}{2\alpha^2}\left(\frac{t}{\beta} + \frac{\beta}{t} - 2\right)\right] & \text{if } t > 0, \end{cases}$$

and the corresponding hazard function for $t > 0$, can be written as

$$h(t; \alpha, \beta) = \frac{f(t; \alpha, \beta)}{1 - \Phi\left[\frac{1}{\alpha}\left\{\left(\frac{t}{\beta}\right)^{1/2} - \left(\frac{\beta}{t}\right)^{1/2}\right\}\right]} = \frac{f(t; \alpha, \beta)}{\Phi\left[\frac{1}{\alpha}\left\{\left(\frac{\beta}{t}\right)^{1/2} - \left(\frac{t}{\beta}\right)^{1/2}\right\}\right]}.$$

The PDF and hazard functions of a Birnbaum-Saunders distribution are unimodal functions for all values of α and β. Since its inception extensive work has been done on different aspects of a Birnbaum-Saunders distribution. The Birnbaum-Saunders distribution and its different variations have been used quite effectively in different areas of science and technology. See for example the recent monograph by Leiva [98] for a comprehensive review of the Birnbaum-Saunders distribution.

Recently Sun and Shi [35] considered the analysis of a simple step-stress model when the lifetime of the experimental units follow two-parameter Birnbaum-Saunders distributions with the same shape parameter α but different scale parameters β_1 and β_2 at the two different stress levels. Hence, based on the CEM assumption the CDF and PDF of the lifetime of an experimental unit from this step-stress experiment become

$$F(t) = \begin{cases} \Phi\left[\frac{1}{\alpha}\left\{\left(\frac{t}{\beta_1}\right)^{1/2} - \left(\frac{t}{\beta_1}\right)^{-1/2}\right\}\right] & \text{if } 0 < t < \tau_1 \\ \Phi\left[\frac{1}{\alpha}\left\{\left(\frac{t-\tau_1}{\beta_2} + \frac{\tau_1}{\beta_1}\right)^{1/2} - \left(\frac{t-\tau_1}{\beta_2} + \frac{\tau_1}{\beta_1}\right)^{-1/2}\right\}\right] & \text{if } \tau_1 \leq t < \infty \end{cases}$$

and

$$f(t) = \begin{cases} g_1(t) & \text{if } 0 < t < \tau_1 \\ g_2(t) & \text{if } \tau_1 \leq t < \infty, \end{cases}$$

where

$$g_1(t) = \frac{1}{2\sqrt{2\pi}\alpha\beta_1}\left[\left(\frac{\beta_1}{t}\right)^{1/2} + \left(\frac{\beta_1}{t}\right)^{3/2}\right]\exp\left[-\frac{1}{2\alpha}\left(\frac{t}{\beta_1} + \frac{\beta_1}{t} - 2\right)\right],$$

$$g_2(t) = \frac{1}{2\sqrt{2\pi}\alpha\beta_2}\left[\left(\frac{t-\tau_1}{\beta_2} + \frac{\tau_1}{\beta_1}\right)^{-1/2} + \left(\frac{t-\tau_1}{\beta_2} + \frac{\tau_1}{\beta_1}\right)^{-3/2}\right]$$
$$\times \exp\left[-\frac{1}{2\alpha}\left\{\left(\frac{t-\tau_1}{\beta_2} + \frac{\tau_1}{\beta_1}\right) + \left(\frac{t-\tau_1}{\beta_2} + \frac{\tau_1}{\beta_1}\right)^{-1} - 2\right\}\right],$$

respectively.

We describe the maximum likelihood estimation procedure based on Type-II censored data. Using the following notations $\xi(t) = t^{1/2} - t^{-1/2}$, $\xi'(t) = (1 - t^{-2})/(2\xi(t))$ and

$$u_i = \begin{cases} \frac{t_{i:n}}{\beta_1} & \text{if} \quad i = 1, 2, \ldots, n_1 \\ \frac{t_{i:n} - \tau_1}{\beta_2} + \frac{\tau_1}{\beta_1} & \text{if} \quad i = n_1 + 1, \ldots, r. \end{cases}$$

the log-likelihood function without the additive constant can be written as

$$l(\alpha, \beta_1, \beta_2) = -n_1 \ln \alpha - n_1 \ln \beta_1 - (r - n_1) \ln \beta_2 + \sum_{i=1}^{r} \ln \xi'(u_i) - \frac{1}{2\alpha^2} \sum_{i=1}^{r} \xi^2(u_i)$$

$$+ (n - r) \ln \left\{ 1 - \Phi \left[\frac{1}{\alpha} \xi(u_r) \right] \right\}; \tag{2.52}$$

see Sun and Shi [35] for details. Therefore, the MLEs of the unknown parameters can be obtained by maximizing Eq. (2.52) with respect to the unknown parameters. The normal equations can be obtained in a standard manner and they need to be solved to compute the MLEs. Sun and Shi [35] proposed to use the asymptotic distribution of the MLEs and bootstrap method to compute confidence intervals of the unknown parameters. Extensive simulation experiments have been performed by Sun and Shi [35], and it is observed that the performances of the MLEs and the associated confidence intervals are not very satisfactory for small sizes, but for moderate and large sample sizes the performances improve.

Since the performances of the MLEs are not very satisfactory, Sun and Shi [35] suggested to use the BEs of the unknown parameters. It is assumed that the scale parameters at the two different stress levels satisfy the power law relationship with the stress, that is

$$\beta_i = \lambda_1 s_i^{\lambda_2}; \quad i = 1, 2.$$

For the power law model, it is assumed that α, λ_1, and λ_2 have the following independent priors:

$$\pi_1(\alpha) \propto \alpha^{-2a-2} \exp(-b\alpha^{-2}), \quad \pi_2(\lambda_1) \propto \frac{1}{\lambda_1}, \quad \pi_3(\lambda_2) \propto \frac{1}{\lambda_2}.$$

Here $a > 0$ and $b > 0$ are the hyperparameters. Based on the previous prior distributions, the posterior distribution for $1 \leq n_1 \leq r - 1$ can be written as

$$\pi(\alpha, \lambda_1, \lambda_1 | \mathcal{D}) \propto \frac{\beta_1^{-n_1} \beta_2^{-(r-n_1)}}{\lambda_1 \lambda_2} \alpha^{-r-2a-2} \prod_{i=1}^{r} \xi'(u_i) \exp \left\{ -\frac{1}{2\alpha^2} \left[\sum_{i=1}^{r} \xi^2(u_i) + 2b \right] \right\}$$

$$\times \left\{ 1 - \Phi \left[\frac{1}{\alpha} \xi(u_r) \right] \right\}^{n-r}.$$

Therefore, the BE of $g(\alpha, \lambda_1, \lambda_2)$, any function of α, λ_1, and λ_2, with respect to squared error loss function can be obtained as the posterior expectation, i.e.,

$$\int_0^\infty \int_0^\infty \int_0^\infty \pi(\alpha, \lambda_1, \lambda_1 | \mathcal{D}) g(\alpha, \lambda_1, \lambda_2) d\alpha d\lambda_1 d\lambda_2, \tag{2.53}$$

provided the expectation exists. It is not possible to compute Eq. (2.53) analytically for general $g(\alpha, \lambda_1, \lambda_2)$. It is possible to approximate Eq. (2.53) by using Lindley's [67] method. However, it is not possible to compute the HPD CRI of $g(\alpha, \lambda_1, \lambda_2)$ using Lindley's method. Sun and Shi [35] used Gibbs sampling technique based on full conditionals mainly following the approaches of Neal [99] and Xu and Tang [100] and obtained the BEs and the associated HPD CRI of $g(\alpha, \lambda_1, \lambda_2)$. Extensive simulation experiments have been performed by the authors and it is observed that the performances of the BEs are better than the MLEs in terms of the biases and MSEs. Moreover, the HPD CRIs also have coverage percentages very close to the corresponding nominal levels.

Open Problems: Prior elicitation is an important problem in any Bayesian inference. Choosing a proper prior is always a challenging task in Bayesian analysis. Recently Wang et al. [101] proposed inverse gamma priors for both the shape and scale parameters for a Birnbaum-Saunders distribution. The performances are quite satisfactory. It will be interesting to see how they perform in case of the Birnbaum-Saunders step-stress model.

2.7 Geometric distribution

So far we have discussed the ALT models when the data are observed in a continuous manner and different continuous distribution functions have been used to analyze these data. Let us now consider the ALT experiment in a discrete set up. Suppose the lifetimes of the experimental units depend on the number of shocks the units receive or the number of times they are switched on and off; then the lifetimes are discrete in nature. In this set-up, if w is the number of shocks the experimental unit receives or the number of times it has been switched on and off before it fails, the integer value w is defined as the failure time of the unit on test. Alternatively, suppose it is not possible to monitor the experiment in a continuous manner and it is being monitored only at the discrete time points; then the number of observations (failures) at any time point will be an integer. Let us assume that the failures are observed only at the discrete time points, say 1, 2, …. Now, n identical units are placed on a life testing experiment at the initial stress level s_1 and the number of and the stress level is increased to s_2 at the prefixed time point τ_1. The problem is to analyze such discrete data obtained from a simple step-stress experiment.

A one-parameter geometric distribution has been used quite extensively to analyze discrete data. The probability mass function (PMF) of a geometric distribution is

$$P(X = x; \theta) = \theta(1 - \theta)^{x-1}; \quad x = 1, 2, \ldots, \tag{2.54}$$

Geometric distribution with the PMF (2.54) will be denoted by GE(θ). The CDF and the hazard function of a GE(θ) are

$$P(X \leq x; \theta) = 1 - (1 - \theta)^{[x]} \text{ for } x \geq 0$$

and

$$h(x; \theta) = \theta, \text{ for } x = 0, 1, \ldots$$

respectively, where $[x]$ denotes the largest integer not exceeding x. It has several interesting properties. It can be considered as the discrete analogue of a one-parameter exponential distribution. It is very similar to the exponential distribution in the sense that it has the lack of memory property and it has a constant hazard function. For different applications and various properties of a geometric distribution, see Johnson et al. [102].

Arefi and Razmkhah [103] first considered the classical inference of the unknown parameters of a simple step-stress model when the lifetime distributions of the experimental units are measured in discrete units and they follow the geometric distribution. The results have been extended for the multiple step-stress model by Wang et al. [58]. Arefi et al. [104] provided the Bayesian inference of the unknown parameters for the simple step-stress model under the same set-up as in Arefi and Razmkhah [103].

Let us assume that the lifetime distribution of the experimental units follow $\text{GE}(\theta_i)$, at the stress level s_i for $i = 1, 2$, and the stress level is changed from s_1 to s_2 at an integer time point τ_1. Here $0 < \theta_1, \theta_2 < 1$. Based on the CEM assumption the CDF of an experimental unit from this step-stress experiment can be written as

$$F(x) = \begin{cases} 1 - (1 - \theta_1)^{[x]} & \text{if } 1 \leq x \leq \tau_1 \\ 1 - (1 - \theta_1)^{\tau_1}(1 - \theta_2)^{[x] - \tau_1} & \text{if } x > \tau_1, \end{cases}$$

where $[x]$ is the largest integer not exceeding x. The corresponding PMF and the hazard function become

$$f(x) = \begin{cases} \theta_1(1 - \theta_1)^{x-1} & \text{if } x = 1, 2, \ldots, \tau_1, \\ \theta_2(1 - \theta_1)^{\tau_1}(1 - \theta_2)^{x - (\tau_1 + 1)} & \text{if } x = \tau_1 + 1, \tau_1 + 2, \ldots, \end{cases}$$

$$h(x) = \begin{cases} \theta_1 & \text{if } x = 1, 2, \ldots, \tau_1, \\ \theta_2 & \text{if } x = \tau_1 + 1, \tau_1 + 2, \ldots, \end{cases}$$

respectively.

Suppose n identical units are put on a life testing experiment at the stress level s_1. At the time point τ_1 (an integer value) the stress level is changed to s_2, and then finally the experiment stops at the time point η (an integer value). Suppose we observe n_1 failures before τ_1 and n_2 failures between τ_1 and η. Hence, in this case the data will be of the form:

$$\mathcal{D} = \left\{ t_{1:n} \leq t_{2:n} \leq \cdots \leq t_{n_1+1:n} \leq \tau_1 < t_{n_1+1:n} \leq \cdots \leq t_{n_1+n_2:n} \leq \eta \right\}.$$

Using the "tie-run" technique of Gan and Bain [105] the likelihood function of θ_1 and θ_2 can be written as

$$L(\theta_1, \theta_2) = c\theta_1^{n_1}(1 - \theta_1)^{D_1}\theta_2^{n_2}(1 - \theta_2)^{D_2}, \tag{2.55}$$

where $c = \frac{n!}{(n-n_1-n_2)!}\left(\prod_{j=1}^{k} z_j!\right)^{-1}$ with k being the number of "tie-runs" with length Z_j for the jth one, and

$$D_1 = \sum_{i=1}^{n_1} t_{i:n} - n_1 + \tau_1(n - n_1),$$

$$D_2 = \sum_{i=n_1+1}^{n_1+n_2} t_{i:n} - (\tau_1 + 1)n_2 + (\eta - \tau_1)(n - n_1 - n_2).$$

Here $0 \leq n_1, n_2, n_1 + n_2 \leq n$, and we use the usual convention that if $m < k$ $\sum_{i=k}^{m} = 0$. In the subsequent sections we provide the MLEs and BEs of the unknown parameters and the associated confidence and CRIs.

2.7.1 Maximum likelihood estimators

The MLEs of the unknown parameters θ_1 and θ_2 can be obtained by maximizing the log-likelihood function (2.55) with respect to θ_1 and θ_2. The log-likelihood function without the additive constant can be written as

$$l(\theta_1, \theta_2) = n_1 \ln\theta_1 + D_1 \ln(1 - \theta_1) + n_2 \ln\theta_2 + D_2 \ln(1 - \theta_2).$$

It is immediate that the MLEs of both θ_1 and θ_2 exist only when $n_1 > 0$ and $n_2 > 0$. Therefore, from now on it is assumed that $n_1 > 0$ and $n_2 > 0$. The normal equations become

$$\dot{l}_{\theta_1} = \frac{n_1}{\theta_1} - \frac{D_1}{1 - \theta_1} = 0,$$

$$\dot{l}_{\theta_2} = \frac{n_2}{\theta_2} - \frac{D_2}{1 - \theta_2} = 0.$$

Hence, the MLEs of θ_1 and θ_2 can be obtained in explicit form as

$$\hat{\theta}_1 = \frac{n_1}{n_1 + D_1} \quad \text{and} \quad \hat{\theta}_2 = \frac{n_2}{n_2 + D_2}.$$

Although, MLEs θ_1 and θ_2 can be obtained in explicit forms, the exact confidence intervals cannot be obtained in explicit forms. Confidence intervals based on the asymptotic distribution of the MLEs can be used to construct asymptotic confidence intervals. In this case the observed Fisher information matrix is as follows:

$$\mathbf{I}_{obs}(\theta_1, \theta_2) = - \begin{bmatrix} \frac{\partial^2 l}{\partial \theta_1^2} & \frac{\partial^2 l}{\partial \theta_1 \partial \theta_2} \\ \frac{\partial^2 l}{\partial \theta_2 \partial \theta_1} & \frac{\partial^2 l}{\partial \theta_2^2} \end{bmatrix} = \begin{bmatrix} \frac{n_1}{\theta_1^2} + \frac{D_1}{(1-\theta_1)^2} & 0 \\ 0 & \frac{n_2}{\theta_2^2} + \frac{D_2}{(1-\theta_2)^2} \end{bmatrix}.$$

Hence, 95% confidence intervals of θ_1 and θ_2 are

$$(\hat{\theta}_1 - 1.96\sqrt{V_1}, \hat{\theta}_1 + 1.96\sqrt{V_1}) \quad \text{and} \quad (\hat{\theta}_2 - 1.96\sqrt{V_2}, \hat{\theta}_2 + 1.96\sqrt{V_2}),$$

respectively, where

$$V_1 = \frac{\theta_1^2(1-\theta_1)^2}{n_1(1-\theta_1)^2 + D_1\theta_1^2} \quad \text{and} \quad V_2 = \frac{\theta_2^2(1-\theta_2)^2}{n_1(1-\theta_2)^2 + D_2\theta_2^2}.$$

Alternatively, the bootstrap method may be used to construct confidence intervals of the unknown parameters in a routine manner.

2.7.2 Bayesian inference

Since, based on the MLEs, it is difficult to obtain the exact confidence intervals of θ_1 and θ_2, Bayesian inference seems to be a reasonable choice particularly for small or moderate sample sizes. Arefi et al. [104] considered the Bayesian inference of the unknown parameters for a geometric simple step-stress model. Since in a general step-stress model as the stress level increases the mean life decreases, therefore it is reasonable to assume that $0 < \theta_1 < \theta_2 < 1$. Arefi et al. [104] assumed Dirichlet prior on (θ_1, θ_2), which has the following PDF:

$$\pi(\theta_1, \theta_2) = \begin{cases} \dfrac{\theta_1^{\lambda \alpha_1 - 1}(\theta_2 - \theta_1)^{\lambda \alpha_2 - 1}(1 - \theta_2)^{\lambda \alpha_3 - 1}}{D(\lambda, \alpha_1, \alpha_2, \alpha_3)} & \text{if} \quad 0 < \theta_1 < \theta_2 < 1 \\ 0 & \text{otherwise,} \end{cases}$$

$$(2.56)$$

where $\lambda > 0$ and $0 < \alpha_1, \alpha_2, \alpha_3 < 1, \alpha_1 + \alpha_2 + \alpha_3 = 1$ are hyperparameters and

$$D(\lambda, \alpha_1, \alpha_2, \alpha_3) = \{\Gamma(\lambda)\}^{-1} \prod_{j=1}^{3} \Gamma(\lambda \alpha_j).$$

The Dirichlet distribution is a very flexible distribution. It has several interesting properties. The generation from a Dirichlet distribution is also very simple. Hence, the simulation experiments and the Bayesian computation can be performed quite conveniently. For a detailed discussion on the Dirichlet distribution see Kotz et al. [106].

The likelihood function (2.55) can be written as follows

$$L(\theta_1, \theta_2) = c \sum_{j=0}^{D_1} \sum_{k=0}^{n_2} \binom{D_1}{j} \binom{n_2}{k} (-1)^{j+k} \theta_1^{n_1+j} (1-\theta_2)^{D_2+k}.$$

Using the Dirichlet prior on θ_1 and θ_2 as defined in Eq. (2.56) the joint posterior distribution of θ_1 and θ_2 can be obtained as

$$\pi(\theta_1, \theta_2 | \mathcal{D}) = \frac{1}{I_0} \sum_{j=0}^{D_1} \sum_{k=0}^{n_2} \left\{ \binom{D_1}{j} \binom{n_2}{k} (-1)^{j+k} \theta_1^{\lambda\alpha_1+n_1+j-1} (\theta_2 - \theta_1)^{\lambda\alpha_2-1} \right.$$

$$\left. \times (1-\theta_2)^{\lambda\alpha_3+D_2+k-1} \right\}, \tag{2.57}$$

where

$$I_0 = \sum_{j=0}^{D_1} \sum_{k=0}^{n_2} \left\{ \binom{D_1}{j} \binom{n_2}{k} (-1)^{j+k} B(\lambda\alpha_3 + D_2 + k, \lambda\alpha_2) \right.$$

$$\left. \times B(\lambda\alpha_1 + n_1 + j, \lambda(1-\alpha_1) + D_2 + k) \right\};$$

here $B(\cdot, \cdot)$ denotes the complete beta function. The marginal posterior distributions of θ_1 and θ_2 can be obtained from Eq. (2.57) as follows:

$$\pi_1(\theta_1 | \mathcal{D}) = \frac{1}{I_0} \sum_{j=0}^{D_1} \sum_{k=0}^{n_2} \left\{ \binom{D_1}{j} \binom{n_2}{k} (-1)^{j+k} B(\lambda\alpha_3 + D_2 + k, \lambda\alpha_2) \right.$$

$$\left. \times \theta_1^{\lambda\alpha_1+n_1+j-1} (1-\theta_1)^{\lambda(1-\alpha_1)+D_2+k-1} \right\} \tag{2.58}$$

and

$$\pi_2(\theta_2 | \mathcal{D}) = \frac{1}{I_0} \sum_{j=0}^{D_1} \sum_{k=0}^{n_2} \left\{ \binom{D_1}{j} \binom{n_2}{k} (-1)^{j+k} B(\lambda\alpha_1 + n_1 + j, \lambda\alpha_2) \right.$$

$$\left. \times \theta_2^{\lambda(1-\alpha_3)+n_1+j-1} (1-\theta_2)^{\lambda\alpha_3+D_2+k-1} \right\}, \tag{2.59}$$

respectively. Therefore, the BEs of θ_1 and θ_2 with respect to the squared error loss functions can be obtained as the corresponding posterior means, and they are as follows:

$$\hat{\theta}_{1B} = E_{\pi_1}(\theta_1|\mathcal{D}) = \frac{1}{I_0} \sum_{j=0}^{D_1} \sum_{k=0}^{n_2} \left\{ \binom{D_1}{j}\binom{n_2}{k}(-1)^{j+k} B(\lambda\alpha_3 + D_2 + k, \lambda\alpha_2) \right.$$

$$\left. \times B(\lambda\alpha_1 + n_1 + j + 1, \lambda(1-\alpha_1) + D_2 + k) \right\},$$

when $\lambda\alpha_1 + n_1 > 2$ and

$$\hat{\theta}_{2B} = E_{\pi_2}(\theta_2|\mathcal{D}) = \frac{1}{I_0} \sum_{j=0}^{D_1} \sum_{k=0}^{n_2} \left\{ \binom{D_1}{j}\binom{n_2}{k}(-1)^{j+k} B(\lambda\alpha_1 + n_1 + j, \lambda\alpha_2) \right.$$

$$\left. \times B(\lambda(1-\alpha_3) + n_1 + j + 1, \lambda\alpha_3 + D_2 + k) \right\},$$

when $(1-\alpha_3)\lambda + n_1 > 2$, respectively.

Although the BEs of θ_1 and θ_1 can be obtained in explicit form, the corresponding CRIs cannot be obtained in explicit form. Let us remember that a $100(1-\alpha)\%$ symmetric CRI, (L, U), of an unknown parameter θ defined on $(0, 1)$ can be obtained from its posterior PDF $\pi(\theta|\mathcal{D})$ by solving the following two nonlinear equations;

$$\frac{\alpha}{2} = \int_0^L \pi(\theta|\mathcal{D})d\theta \quad \text{and} \quad \frac{\alpha}{2} = \int_U^1 \pi(\theta|\mathcal{D})d\theta.$$

It can be shown using Eqs. (2.58), (2.59) that a $100(1-\alpha)\%$ symmetric CRI of θ_i, for $i = 1, 2$, is the solution of the following two nonlinear equations

$$\frac{\alpha}{2} = \psi_i(L_i) \quad \text{and} \quad 1 - \frac{\alpha}{2} = \psi_i(U_i), \tag{2.60}$$

where L_i and U_i are the lower and upper limits of the associated symmetric CRI, respectively, and

$$\psi_1(z) = \frac{1}{I_0} \sum_{j=0}^{D_1} \sum_{k=0}^{n_2} \left\{ \binom{D_1}{j}\binom{n_2}{k}(-1)^{j+k} B(\lambda\alpha_3 + D_2 + k, \lambda\alpha_2) \right.$$

$$\left. \times B_{ic}\{z; \lambda\alpha_1 + n_1 + j, \lambda(1-\alpha_1) + D_2 + k\} \right\},$$

$$\psi_2(z) = \frac{1}{I_0} \sum_{j=0}^{D_1} \sum_{k=0}^{n_2} \left\{ \binom{D_1}{j}\binom{n_2}{k}(-1)^{j+k} B(\lambda\alpha_1 + n_1 + j, \lambda\alpha_2) \right.$$

$$\left. \times B_{ic}\{z; \lambda(1-\alpha_3) + n_1 + j, \lambda\alpha_3 + D_2 + k\} \right\},$$

and

$$B_{ic}(z; a, b) = \int_0^z t^{a-1}(1-t)^{b-1}dt,$$

is the incomplete beta function. Hence, to compute symmetric CRI one needs to solve the two nonlinear equations (2.60) by using some iterative methods.

2.8 Multiple step-stress model

So far we have discussed the simple step-stress model and its inference procedure based on different lifetime distributions. In this section we discuss the multiple step-stress model and the associated inference. It may be mentioned that although extensive work has been done on simple step-stress models, the number of references related to multiple step-stress models is rather limited, mainly because of the analytical intractability. Most of the work related to the multiple step-stress models is based on the exponential distribution. In this section we mainly consider the analysis of the multiple step-stress model when the lifetime distribution of the experimental units follows the one-parameter exponential distribution with different scale parameters at different stress levels and it satisfies CEM assumptions.

In a multiple step-stress experiment all the experimental units are subjected to the stress level s_1 at the initial stage. It continues till the time point τ_1 when the stress level is increased to s_2 and so on. Finally at the time point τ_{k-1} the stress level is increased to s_k. The experiment continues till all the experimental units fail. If F_j denotes the CDF of the lifetime of an item at the stress level s_i, for $i = 1, \ldots, k$, then based on the CEM assumption the CDF of a multiple step-stress model can be written as

$$F(t) = \begin{cases} 0 & \text{if } t < 0 \\ F_1(t) & \text{if } 0 \leq t < \tau_1 \\ F_2(t - \tau_1 + h_1) & \text{if } \tau_1 \leq t < \tau_2 \\ F_3(t - \tau_2 + h_2) & \text{if } \tau_2 \leq t < \tau_3 \\ \vdots & \vdots \quad \vdots \\ F_{k-1}(t - \tau_{k-2} + h_{k-2}) & \text{if } \tau_{k-2} \leq t < \tau_{k-1} \\ F_k(t - \tau_{k-1} + h_{k-1}) & \text{if } \tau_{k-1} \leq t < \infty; \end{cases} \qquad (2.61)$$

here h_i's are the solutions of the following sets of equations:

$$F_{i+1}(h_i) = F_i(\tau_i - \tau_{i-1} + h_{i-1}); \quad i = 1, 2, \ldots, k - 1,$$

and $h_0 = \tau_0 = 0$. See for example Gouno and Balakrishnan [107]. Now to develop any inferential procedures, specific assumptions on F_j's need to be made.

Suppose that the lifetime at the stress level s_j follows an exponential distribution with mean θ_j, for $j = 1, 2, \ldots, k$, then Eq. (2.61) can be written as

$$
F(t) = \begin{cases}
0 & \text{if } t < 0 \\
1 - \exp\left\{-\frac{t}{\theta_1}\right\} & \text{if } 0 \le t < \tau_1 \\
1 - \exp\left\{-\frac{1}{\theta_2}(t - \tau_1 + h_1)\right\} & \text{if } \tau_1 \le t < \tau_2 \\
1 - \exp\left\{-\frac{1}{\theta_3}(t - \tau_2 + h_2)\right\} & \text{if } \tau_2 \le t < \tau_3 \\
\vdots & \vdots \quad \vdots \\
1 - \exp\left\{-\frac{1}{\theta_{k-1}}(t - \tau_{k-2} + h_{k-2})\right\} & \text{if } \tau_{k-2} \le t < \tau_{k-1} \\
1 - \exp\left\{-\frac{1}{\theta_k}(t - \tau_{k-1} + h_{k-1})\right\} & \text{if } \tau_{k-1} \le t < \infty,
\end{cases}
\tag{2.62}
$$

and h_i's can be obtained by solving

$$
1 - \exp\left\{-\frac{h_i}{\theta_{i+1}}\right\} = 1 - \exp\left\{-\frac{\tau_i - \tau_{i-1} + h_{i-1}}{\theta_i}\right\}; \quad i = 1, 2, \ldots, k - 1,
$$

where $h_0 = \tau_0 = 0$. Hence, we obtain

$$
h_{m-1} = \theta_m \sum_{j=1}^{m-1} \frac{\tau_j - \tau_{j-1}}{\theta_j}; \quad m = 1, 2, \ldots, k.
$$

The PDF associated with Eq. (2.62) can be written as

$$
f(t) = \begin{cases}
0 & \text{if } t < 0 \\
\frac{1}{\theta_1} \exp\left\{-\frac{t}{\theta_1}\right\} & \text{if } 0 \le t < \tau_1 \\
\frac{1}{\theta_2} \exp\left\{-\frac{t-\tau_1}{\theta_2} - \frac{\tau_1}{\theta_1}\right\} & \text{if } \tau_1 \le t < \tau_2 \\
\frac{1}{\theta_3} \exp\left\{-\frac{t-\tau_2}{\theta_3} - \frac{\tau_2-\tau_1}{\theta_2} - \frac{\tau_1}{\theta_1}\right\} & \text{if } \tau_2 \le t < \tau_3 \\
\vdots & \vdots \quad \vdots \\
\frac{1}{\theta_k} \exp\left\{-\frac{t-\tau_{k-1}}{\theta_k}\right. & \\
\left. -\frac{\tau_{k-1}-\tau_{k-2}}{\theta_{k-1}} - \cdots - \frac{\tau_1}{\theta_1}\right\} & \text{if } \tau_{k-1} \le t < \infty.
\end{cases}
\tag{2.63}
$$

The problem is to estimate the unknown parameters namely $\theta_1, \ldots, \theta_k$ based on the observations from a multiple step-stress experiment. The form of the data is assumed as follows:

$$
t_{11} < \cdots < t_{1n_1} < \tau_1 < t_{21} < \cdots < t_{2n_2} < \tau_2 < \cdots < t_{(k-1)n_{k-1}}
$$
$$
< \tau_{k-1} < \cdots < t_{kn_k}.
\tag{2.64}
$$

This data set indicates that we have n_i observations at the s_ith stress level for $i = 1, 2, \ldots, k$, and $n_1 + n_2 + \cdots + n_k = n$.

Based on the PDF (2.63) the likelihood function of the observed data (Eq. 2.64) can be written as

$$L(\theta_1, \ldots, \theta_k) = \left(\prod_{i=1}^{k} \frac{1}{\theta_i^{n_i}} \right) \exp \left(- \sum_{i=1}^{k} \frac{U_i}{\theta_i} \right),$$

where $\tilde{n}_i = \sum_{j=1}^{i} n_j$ and $U_i = \sum_{j=1}^{n_i} (t_{ij} - \tau_{i-1}) + (n - \tilde{n}_i)(\tau_i - \tau_{i-1})$ for $i = 1, \ldots, k$. It is immediate that if $n_j = 0$, the MLE of θ_j, for $j = 1, \ldots, k$, does not exist. Hence for large k, the probability of nonexistence of the MLE can be quite high.

For this reason often it is assumed that the scale parameters satisfy a log-linear link function with the corresponding stress factor. Mathematically it can be represented as follows:

$$\ln \theta_i = \alpha + \beta s_i; \quad i = 1, \ldots, k. \tag{2.65}$$

Based on the relation (2.65), the problem reduces to estimating the unknown parameters α and β based on the observed data (Eq. 2.64). Based on the assumption (2.65), the likelihood function (2.66) as a function of α and β can be written as

$$L(\alpha, \beta) = \prod_{i=1}^{k} \frac{1}{e^{n_i(\alpha + \beta s_i)}} \exp \left(- \sum_{i=1}^{k} U_i e^{-(\alpha + \beta s_i)} \right). \tag{2.66}$$

Hence, the log-likelihood function becomes

$$l(\alpha, \beta) = -\alpha n - \beta \sum_{i=1}^{k} n_i s_i - e^{-\alpha} \sum_{i=1}^{k} U_i e^{-\beta s_i}. \tag{2.67}$$

Therefore, the MLEs of α and β can be obtained by maximizing Eq. (2.67) with respect to the unknown parameters. They can be obtained by solving the following two normal equations simultaneously.

$$\dot{l}_\alpha = -n + e^{-\alpha} \sum_{i=1}^{k} U_i e^{-\beta s_i} = 0, \tag{2.68}$$

$$\dot{l}_\beta = -\beta \sum_{i=1}^{k} n_i s_i + e^{-\alpha} \sum_{i=1}^{k} s_i U_i e^{-\beta s_i} = 0. \tag{2.69}$$

Clearly, the two normal equations (2.68) and (2.69) cannot be solved analytically. One needs to use some numerical algorithm like the Newton-Raphson method to solve them. Alternatively, by maximizing the profile log-likelihood function, the MLEs of α and β can be obtained as follows.

First observe that for a given β, the MLEs of α can be obtained from Eq. (2.68) as

$$\hat{\alpha}(\beta) = \ln\left(\frac{1}{n}\sum_{i=1}^{k}U_i e^{-\beta_i s_i}\right).$$

Now substituting $\hat{\alpha}(\beta)$ in Eq. (2.67) and ignoring the additive constant, we obtain the profile log-likelihood function of β as

$$g(\beta) = l(\hat{\alpha}(\beta),\beta) = -n\ln\left(\frac{1}{n}\sum_{i=1}^{k}U_i e^{-\beta_i s_i}\right) - \beta\sum_{i=1}^{k}n_i s_i. \tag{2.70}$$

Therefore, the MLE of β can be obtained by maximizing Eq. (2.70) with respect to β and if we denote it as $\hat{\beta}$, then the MLE of α can be obtained as $\hat{\alpha} = \hat{\alpha}(\hat{\beta})$.

It is not possible to obtain the exact distribution of $\hat{\alpha}$ and $\hat{\beta}$, and because of that bootstrap confidence intervals can be used for construction of the confidence intervals of the unknown parameters. Alternatively, the approximate confidence intervals based on the asymptotic distributions of $\hat{\alpha}$ and $\hat{\beta}$ can also be used in this case.

As we have mentioned before, although extensive work has been done on a simple step-stress model the amount of work related to the multiple step-stress model is rather limited. There are some issues related to the multiple step-stress model. For example, designing a multiple step-stress model and the ordered restricted inference of the model parameters are important problems, and we will discuss them in the subsequent chapters.

Other related models

3

3.1 Introduction

In the previous chapter we have discussed the development of inferential issues of the step-stress life test (SSLT) under the assumptions of the cumulative exposure model (CEM). In this chapter we will discuss three other popular models, viz., the tempered random variable model (TRVM), tempered failure rate model (TFRM), and cumulative risk model (CRM). As we have provided detail introductions of these models in Section 1.5, here we will briefly mention the models and proceed to the inferential issues for different parametric models under various censoring schemes.

The rest of the chapter is organized as follows. In Section 3.2, we discuss the inferential issues of the TRVM. The issues of the estimation of unknown parameters under the assumptions of TFRM are discussed in Section 3.3. We will address the CRM and associated inferential procedures in Section 3.4.

3.2 Tempered random variable model

As discussed in Section 1.5.2, the TRVM assumes that the effect of the change of the stress level from s_1 to s_2 at the time τ_1 is equivalent to multiplying the remaining life of the unit by an unknown positive constant β, which depends on both the stress levels. Let T and $\widetilde{T}$ denote the lifetime of a test item under the stress level s_1 and the total lifetime of the test item under the step-stress pattern, respectively. Mathematically, T and $\widetilde{T}$ are related by

$$\widetilde{T} = \begin{cases} T & \text{if } T \le \tau_1 \\ \tau_1 + \frac{T - \tau_1}{\beta} & \text{if } T > \tau_1. \end{cases} \tag{3.1}$$

Since the effect of switching to the higher stress level is to shorten the lifetime of the test unit, usually β is greater than one. Furthermore, unlike the CEM, the CDFs of the lifetime under different stress levels are related to each other by Eq. (3.1) for the TRVM. The time τ_1 is called the tampering time. If the observed lifetime of an item is greater than τ_1, the observation is called a tampered observation. Otherwise, it is called an untampered observation.

Analysis of Step-Stress Models. http://dx.doi.org/10.1016/B978-0-12-809713-7.00003-X

The inferential issues have been addressed by several authors under the assumption of TRVM for different distributions of the random variable T and for different censoring schemes. Goel [17] considered the distribution of T to be uniform over the interval (0, 1) and showed the consistency of the maximum likelihood estimators (MLEs) under mild conditions. Based on the assumption of a one-parameter exponential distribution on T, Goel [18] showed that the MLEs of unknown parameters are strongly consistent and asymptotically normally distributed. DeGroot and Goel [19] considered the Bayesian estimation of the unknown model parameters of the same model under a gamma prior. They also addressed the optimal choice of the tampering points for different loss functions under the Bayesian framework. The optimal choice of the tampering point was considered by Bai and Chung [108] for the one-parameter exponential distribution under the classical set-up. Abdel-Ghaly et al. [109] considered the likelihood estimation of model parameters in the presence of Type-I and Type-II censoring under the assumption that T has a Weibull distribution.

3.2.1 One-parameter exponential distribution

In this subsection, we discuss the inferential issues of the model parameters when the lifetime of an experimental unit under the first stress level has a one-parameter exponential distribution with mean $\frac{1}{\lambda}$ having the following PDF:

$$f_T(t) = \begin{cases} \lambda e^{-\lambda t} & \text{if } t \geq 0 \\ 0 & \text{otherwise,} \end{cases} \tag{3.2}$$

where $\lambda > 0$ is a real number. Using Eq. (1.4), the PDF of the total lifetime, $\widetilde{T}$, of an experimental unit under the step-stress pattern is given by

$$f_{\widetilde{T}}(t) = \begin{cases} \lambda e^{-\lambda t} & \text{if } t \leq \tau_1 \\ \beta \lambda e^{-\lambda(\tau_1 + \beta(t - \tau_1))} & \text{if } t > \tau_1 \\ 0 & \text{otherwise.} \end{cases} \tag{3.3}$$

Suppose n items are put on a life testing experiment at the stress level s_1. The stress level is increased from s_1 to s_2 at a prefixed time τ_1. The test is terminated at a prespecified time $\eta > \tau_1$. Bai and Chung [108] assumed that the distribution of the lifetime under the step-stress pattern is given by Eq. (3.3). Recall that the ordered observed data under a Type-I censoring scheme can have one of the following forms:

(a) $\tau_1 < t_{1:n} < \cdots < t_{n_2:n} < \eta$,
(b) $t_{1:n} < \cdots < t_{n_1:n} < \tau_1 < t_{n_1+1:n} < \cdots < t_{n_1+n_2:n} < \eta$,
(c) $t_{1:n} < \cdots < t_{n_1:n} < \tau_1 < \eta$.

The likelihood function of the observed data can be expressed as

$$L(\lambda, \beta) = \lambda^{n_1+n_2} \beta^{n_2} e^{-\lambda\left(\sum_{i=1}^{n_1} t_{i:n} + (n-n_1)\tau_1 + \beta \sum_{i=n_1+1}^{n_2} (t_{i:n} - \tau_1) + \beta(n-n_1-n_2)(\eta-\tau_1)\right)}.$$

The MLEs of λ and β do not exist if $n_1 = 0$ or $n_2 = 0$. If $n_1 \neq 0$ and $n_2 \neq 0$, the MLEs of the unknown parameters exist and are unique. They are given by

$$\hat{\lambda} = \frac{n_1}{A_1} \quad \text{and} \quad \hat{\beta} = \frac{n_2 A_1}{n_1 B_1},$$

where $A_1 = \sum_{i=1}^{n_1} t_{i:n} + (n-n_1)\tau_1$ and $B_1 = (n-n_1-n_2)(\eta-\tau_1) + \sum_{i=n_1+1}^{n_2} (t_{i:n} - \tau_1)$. Clearly, $\hat{\lambda}$ and $\hat{\beta}$ are the conditional MLEs of λ and β conditioned on the event that there is at least one failure at each of the stress levels. The Fisher information matrix is given by

$$F(\lambda, \beta) = \begin{bmatrix} F_{11} & F_{12} \\ F_{21} & F_{22} \end{bmatrix},$$

where $F_{11} = \frac{n(p_1+p_2)}{\lambda^2}$, $F_{12} = F_{21} = \frac{np_1}{\lambda\beta}$, $F_{22} = \frac{np_2}{\beta^2}$, $p_1 = 1 - e^{-\lambda\tau_1}$, and $p_2 = e^{-\lambda\tau_1}\left(1 - e^{-\lambda\beta(\eta-\tau_1)}\right)$. Bai and Chung [108] obtained the optimal stress changing time τ_1 for the given values of λ, β, and η. We will discuss the optimality issues in Chapter 5.

The Bayesian estimation of λ and β is considered by DeGroot and Goel [19] under the assumption that different items have different tampering times and the complete data are available. Here we highlight the results of DeGroot and Goel [19] for the simple SSLT, that is the tampering times are the same for all the items. The authors separately considered the estimation problem of β when λ is known and the estimation problem of λ and β when both the parameters are unknown. When λ is known, it is assumed that the prior distribution on β is a Gamma$(v, s\lambda)$, where $v > 0$ and $s > 0$. As the gamma distribution belongs to a conjugate family, the posterior distribution of β given the data is a Gamma$(v_1, s_1\lambda)$, where $v_1 = v + n_2$ and $s_1 = s + \sum_{i=n_1+1}^{n} (t_{i:n} - \tau_1)$. The posterior PDF of β is given by

$$\pi(\beta|\text{Data}) \propto \beta^{v_1-1} e^{-s_1\lambda\beta} \quad \text{for } \beta > 0.$$

The loss function used for the estimation of β is given by

$$L^*(\hat{\beta}, \beta) = \hat{\beta}^k \beta^l (\hat{\beta} - \beta)^2, \tag{3.4}$$

where $-2 \leq k \leq 0$ and $-\infty < l < \infty$. As the Bayes risk is infinite for $k < -2$ and zero for $k > 0$, the values of k in the interval $[-2, 0]$ are considered. For $k = 0$ and $l = -2$, the loss function (3.4) becomes $L_1(\hat{\beta}, \beta) = \left(\frac{\hat{\beta}}{\beta} - 1\right)^2$, which measures the squared error relative to the true value of β. Again for $k = -2$ and $l = 0$, the loss function (3.4) becomes $L_2(\hat{\beta}, \beta) = \left(\frac{\beta}{\hat{\beta}} - 1\right)^2$, which measures the squared error relative to the estimate of β. Note that β is a scale parameter and both the loss functions L_1 and L_2 are invariant under a constant multiplication to all the data points. Now, we will state a theorem regarding the BE of β under the loss function L^* as defined in

Eq. (3.4). Interested readers can find the proof of the theorem in DeGroot and Goel [19] for a general set-up.

Theorem 3.2.1. *For $v+l > 0$, the BE of β with respect to the loss function $L^*(\hat{\beta}, \beta)$, as defined in Eq. (3.4), is given by*

$$\hat{\beta} = \gamma_k(\eta_1)\frac{\eta_1}{s_1\lambda},$$

where $\eta_1 = v + n - n_1 + l$ and for $x > 0$

$$\gamma_k(x) = \begin{cases} \frac{1}{k+2}\left(k+1+\sqrt{1-k(k+2)x^{-1}}\right) & \text{if } -2 < k \leq 0 \\ 1+x^{-1} & \text{if } k = -2. \end{cases}$$

When λ is unknown, one can consider a conjugate prior on λ and β as follows. Given $\lambda = \lambda_0$, the conditional prior distribution on β is a Gamma(v, $s\lambda_0$), and the prior distribution on λ is a Gamma(v_0, s_0). The joint prior PDF of λ and p is given by

$$\pi(\lambda, \beta) = \frac{s^v s_0^{v_0}}{\Gamma(v)\Gamma(v_0)}\beta^{v-1}\lambda^{v+v_0-1}e^{-\lambda(s_0+s\beta)} \quad \text{for} \quad \lambda > 0, \beta > 0,$$

and zero otherwise. The posterior PDF of λ and β is

$$\pi(\lambda, \beta|\text{Data}) \propto \lambda^{v+v_0+n-1}\beta^{v+n-n_1-1}e^{-\lambda(s_2+\beta s_1)} \quad \text{for} \quad \lambda > 0, \beta > 0,$$

and zero otherwise, where $s_2 = s_0 + \sum_{i=1}^{n_1} t_{i:n} + (n - n_1)\tau_1$. Now, we will state two theorems regarding the BEs of λ and β under the loss function L^* as defined in Eq. (3.4). Interested readers can find the proofs in DeGroot and Goel [19] for a more general set-up.

Theorem 3.2.2. *If $v_0 + l > 2$, the BE of λ with respect to the loss function L^*, as defined in Eq. (3.4), is given by*

$$\hat{\lambda} = \gamma_k(\eta_2)\frac{\eta_2}{s_2},$$

where $\eta_2 = v_0 + n_1 + l$ and $\gamma_k(\cdot)$ is defined in Theorem 3.2.1.

Theorem 3.2.3. *If $v + l > 0$ and $v_0 - l > 2$, the BE of β with respect to the loss function L^*, as defined in Eq. (3.4), is given by*

$$\hat{\beta} = \gamma_k(\eta_3)\frac{\eta_4 s_2}{s_1},$$

where $\eta_3 = \frac{(v+n-n_1+l)(v_0+n_1-l-2)}{v_0+v+n-1}$, $\eta_4 = \frac{v+n-n_1+l}{v_0+n_1-l-1}$, and $\gamma_k(\cdot)$ is defined in Theorem 3.2.1.

DeGroot and Goel [19] assumed that the loss function for the simultaneous estimation of λ and β is a linear combination of the losses resulting from the estimation of each of the parameters separately; therefore the authors considered the following loss function for the simultaneous estimation of λ and β:

$$L_3(\hat{\lambda}, \hat{\beta}, \lambda, \beta) = \rho_1 \hat{\lambda}^{k_1} \lambda^{l_1} (\hat{\lambda} - \lambda)^2 + \rho_2 \hat{\beta}^{k_2} \beta^{l_2} (\hat{\beta} - \beta)^2,$$

where $-2 \leq k_1 \leq 0, -\infty < l_1 < \infty, -2 \leq k_2 \leq 0, -\infty < l_1 < \infty, \rho_1 > 0,$ and $\rho_2 > 0$. The BEs of λ and β with respect to the loss function L_3 are given by Theorems 3.2.2 and 3.2.3 with the appropriate choices of k and l.

Now, let us consider the following open problems for future work.

Open Problem: Bai and Chung [108] addressed the classical inference of a simple SSLT under the assumptions of TRVM and exponentially distributed lifetimes when the lifetimes are Type-I censored. The authors obtained the MLEs of the model parameters. However, they have not considered the exact distributions of the MLEs of the unknown parameters. It would be interesting to find the exact distribution of the MLEs and then use them to find the interval estimators of the model parameters.

Open Problem: Under the same set-up as Bai and Chung [108], one may want to establish the consistency and asymptotic normality of the MLEs of the model parameters. This problem will be quite interesting due to the presence of Type-I censoring scheme. It may be mentioned that Goel [18] considered the complete sample case for the one-parameter exponential distribution.

Open Problem: The Bayesian inference of the simple SSLT was developed by DeGroot and Goel [19] under the assumptions of the TRVM. The authors assumed that the lifetime has a one-parameter exponential distribution and the complete data are available. However, in practice most of the data are censored. Hence, it would be more appealing to extend the work of DeGroot and Goel [19] for different censoring schemes.

3.2.2 Weibull distribution

In this section we will assume that the distribution of lifetime at the first stress level is Wei($\alpha, \frac{1}{\theta^\alpha}$), i.e., the PDF of T, the lifetime at the first stress level, is given by

$$f_T(t) = \frac{\alpha}{\theta} \left(\frac{t}{\theta}\right)^{\alpha-1} e^{-(t/\theta)^\alpha} \quad \text{for} \quad t > 0,$$

and zero otherwise. Therefore using Eq. (1.4), the PDF of $\widetilde{T}$, the lifetime under the step-stress pattern can be expressed as

$$f_{\widetilde{T}}(t) = \begin{cases} f_1(t) & \text{if } 0 < t \leq \tau_1 \\ f_2(t) & \text{if } t > \tau_1 \\ 0 & \text{otherwise,} \end{cases} \tag{3.5}$$

where

$$f_1(t) = \frac{\alpha}{\theta} \left(\frac{t}{\theta}\right)^{\alpha-1} e^{-(t/\theta)^\alpha},$$

$$f_2(t) = \frac{\beta\alpha}{\theta} \left(\frac{\tau_1 + \beta(t - \tau_1)}{\theta}\right)^{\alpha-1} e^{-((\tau_1+\beta(t-\tau_1))/\theta)^\alpha}.$$

Abdel-Ghaly et al. [109] developed the classical inference of the model parameters in the presence of Type-I and Type-II censoring schemes. They assumed that $\tilde{T}$ has a PDF defined in Eq. (3.5). Now we will describe the inferential procedure in detail.

Analysis of Type-I censored data

Let the experiment be terminated at prefixed time $\eta \, (> \tau_1)$. The form of the data is given in the previous subsection. The likelihood function of the observed data is given by

$$
L(\alpha, \theta, \beta) \propto \begin{cases} \prod_{i=1}^{n_2} f_2(t_{i:n}) \, (1 - F_2(\eta))^{n-n_2} & \text{in case (a)} \\ \prod_{i=1}^{n_1} f_1(t_{i:n}) \prod_{i=n_1+1}^{n_1+n_2} f_2(t_{i:n}) \, (1 - F_2(\eta))^{n-n_1-n_2} & \text{in case (b)} \\ \prod_{i=1}^{n_1} f_1(t_{i:n}) \, (1 - F_2(\eta))^{n-n_1} & \text{in case (c),} \end{cases}
$$

where $F_2(\eta) = 1 - e^{-((\tau_1+\beta(\eta-\tau_1))/\theta)^\alpha}$. Assuming that $n_1 = 0$ in case (a) and $n_2 = 0$ in case (b), the log-likelihood function, ignoring the constant term, can be written as

$$
\begin{aligned}
l(\alpha, \theta, \beta) = &(n_1 + n_2) \ln \alpha + n_2 \ln \beta - (n_1 + n_2)\alpha \ln \theta \\
&+ (\alpha - 1) \left(\sum_{i=1}^{n_1} \ln t_{i:n} + \sum_{i=n_1+1}^{n_1+n_2} \ln (\tau_1 + \beta(t_{i:n} - \tau_1)) \right) - \frac{Q(\alpha, \beta)}{\theta^\alpha},
\end{aligned}
\tag{3.6}
$$

where $Q(\alpha, \beta) = \sum_{i=1}^{n_1} t_{i:n}^\alpha + \sum_{i=n_1+1}^{n_1+n_2} (\tau_1 + \beta(t_{i:n} - \tau_1))^\alpha + (n - n_1 - n_2) (\tau_1 + \beta(\eta - \tau_1))^\alpha$. The MLEs of the unknown parameters can be found by maximizing Eq. (3.6) with respect to α, β, and θ. The authors did not discuss the issue of the existence of MLEs of the parameters. It seems that the MLEs of α, θ, and β exist if $n_2 \neq 0$ and $n_1 + n_2 \geq 3$. However, explicit proof is needed to establish this fact. Taking first order partial derivatives with respect to α, β, and θ, and equating them to zero, one can write the following normal equations:

$$
\begin{aligned}
\dot{l}_\alpha = &(n_1 + n_2) \left(\frac{1}{\alpha} - \ln \theta \right) + \sum_{i=1}^{n_1} \ln t_{i:n} \\
&+ \sum_{i=n_1+1}^{n_1+n_2} \ln (\tau_1 + \beta(t_{i:n} - \tau_1)) - \frac{\dot{Q}_\alpha - Q \ln \theta}{\theta^\alpha} = 0,
\end{aligned}
\tag{3.7}
$$

$$
\dot{l}_\beta = \frac{n_2}{\beta} + (\alpha - 1) \sum_{i=n_1+1}^{n_1+n_2} \frac{t_{i:n} - \tau_1}{\tau_1 + \beta(t_{i:n} - \tau_1)} - \frac{\dot{Q}_\beta}{\theta^\alpha} = 0,
\tag{3.8}
$$

$$
\dot{l}_\theta = -(n_1 + n_2) \frac{\alpha}{\theta} + \frac{\alpha Q}{\theta^{\alpha+1}} = 0,
\tag{3.9}
$$

where

$$Q_\beta = \alpha \sum_{i=n_1+1}^{n_1+n_2} (t_{i:n} - \tau_1)(\tau_1 + \beta(t_{i:n} - \tau_1))^{\alpha-1}$$

$$+ \alpha(n - n_1 - n_2)(\eta - \tau_1)(\tau_1 + \beta(\eta - \tau_1))^{\alpha-1},$$

$$Q_\alpha = \sum_{i=1}^{n_1} t_{i:n}^\alpha \ln t_{i:n} + \sum_{i=n_1+1}^{n_1+n_2} (\tau_1 + \beta(t_{i:n} - \tau_1))^\alpha \ln(\tau_1 + \beta(t_{i:n} - \tau_1))$$

$$+ (n - n_1 - n_2)(\tau_1 + \beta(\eta - \tau_1))^\alpha \ln(\tau_1 + \beta(\eta - \tau_1)).$$

From Eq. (3.9), for fixed values of α and β, the MLE of θ can be expressed as

$$\hat\theta(\alpha, \beta) = \left(\frac{Q(\alpha, \beta)}{n_1 + n_2} \right)^{1/\alpha}. \tag{3.10}$$

Substituting θ by $\hat\theta(\alpha, \beta)$ in Eqs. (3.7), (3.8), the problem is reduced into solving two nonlinear equations given by

$$\frac{n_2}{\beta} + (\alpha - 1) \sum_{i=n_1+1}^{n_1+n_2} \frac{t_{i:n} - \tau_1}{\tau_1 + \beta(t_{i:n} - \tau_1)} - (n_1 + n_2) \frac{Q_\beta}{Q(\alpha, \beta)} = 0, \tag{3.11}$$

$$\frac{n_1 + n_2}{\alpha} + \sum_{i=1}^{n_1} \ln t_{i:n} + \sum_{i=n_1+1}^{n_1+n_2} \ln(\tau_1 + \beta(t_{i:n} - \tau_1)) - (n_1 + n_2) \frac{Q_\alpha}{Q(\alpha, \beta)} = 0. \tag{3.12}$$

Since the closed forms of solutions to Eqs. (3.11), (3.12) are difficult to obtain, one needs to use some numerical technique to solve them. Abdel-Ghaly et al. [109] suggested using the modified quasilinear method which has been proposed by Miele et al. [110]. Once $\hat\alpha$ and $\hat\beta$, the MLEs of α and β, respectively, are obtained, they can be substituted in Eq. (3.10) to get the MLE of θ. One of the crucial issues of the numerical procedure is to choose the proper initial values. The natural choice, which may not be the best one, is to start with the exponential case, i.e., $\alpha = 1$. For all values of $x > a > 0$, $P(X \le x) = P(X - a \le x | X > a)$, when X has a one-parameter exponential distribution. Hence, a plausible initial value of β might be

$$\frac{\frac{1}{n_1} \sum_{i=1}^{n_1} t_{i:n}}{\frac{1}{n-n_1} \sum_{i=n_1+1}^{n} t_{i:n} - \tau_1}.$$

Analysis of Type-II censored data

The data obtained from a simple SSLT under a Type-II censoring scheme will have one of the following forms:

(a) $\tau_1 < t_{1:n} < \cdots < t_{r:n}$,
(b) $t_{1:n} < \cdots < t_{n_1:n} < \tau_1 < t_{n_1+1:n} < \cdots < t_{r:n}; 0 < n_1 < r$,
(c) $t_{1:n} < \cdots < t_{r:n} < \tau_1$.

Writing $n_1 = 0$ in case (a) and $n_1 = r$ in case (b), the log likelihood function of α, β, and θ, ignoring the constant terms, can be expressed as

$$l(\alpha, \theta, \beta) = r \ln \alpha - (r - n_1) \ln \beta - r\alpha \ln \theta - \frac{W(\alpha, \beta)}{\theta^\alpha}$$

$$+ (\alpha - 1) \left(\sum_{i=1}^{n_1} \ln t_{i:n} + \sum_{i=n_1+1}^{r} \ln (\tau_1 + \beta(t_{i:n} - \tau_1)) \right), \qquad (3.13)$$

where

$$W(\alpha, \beta) = \sum_{i=1}^{n_1} t_{i:n}^\alpha + \sum_{i=n_1+1}^{r} (\tau_1 + \beta(t_{i:n} - \tau_1))^\alpha + (n - r)(\tau_1 + \beta(t_{r:n} - \tau_1))^\alpha.$$

The MLEs can be found by maximizing Eq. (3.13) with respect to α, β, and θ. Taking the first order partial derivatives with respect to α, β, and θ, and equating them to zero, one can write the following normal equations:

$$\dot{l}_\alpha = r \left(\frac{1}{\alpha} - \ln \theta \right) + \sum_{i=1}^{n_1} \ln t_{i:n} + \sum_{i=n_1+1}^{r} \ln (\tau_1 + \beta(t_{i:n} - \tau_1)) - \frac{\dot{W}_\alpha - W \ln \theta}{\theta^\alpha} = 0,$$
$$(3.14)$$

$$\dot{l}_\beta = \frac{r - n_1}{\beta} + (\alpha - 1) \sum_{i=n_1+1}^{r} \frac{t_{i:n} - \tau_1}{\tau_1 + \beta(t_{i:n} - \tau_1)} - \frac{\dot{W}_\beta}{\theta^\alpha} = 0, \qquad (3.15)$$

$$\dot{l}_\theta = -r \frac{\alpha}{\theta} + \frac{\alpha W}{\theta^{\alpha+1}} = 0, \qquad (3.16)$$

where

$$\dot{W}_\beta = \alpha \sum_{i=n_1+1}^{r} (t_{i:n} - \tau_1)(\tau_1 + \beta(t_{i:n} - \tau_1))^{\alpha-1}$$

$$+ \alpha(n - r)(t_{r:n} - \tau_1)(\tau_1 + \beta(t_{r:n} - \tau_1))^{\alpha-1},$$

$$\dot{W}_\alpha = \sum_{i=1}^{n_1} t_{i:n}^\alpha \ln t_{i:n} + \sum_{i=n_1+1}^{r} (\tau_1 + \beta(t_{i:n} - \tau_1))^\alpha \ln (\tau_1 + \beta(t_{i:n} - \tau_1))$$

$$+ (n - r)(\tau_1 + \beta(t_{r:n} - \tau_1))^\alpha \ln (\tau_1 + \beta(t_{r:n} - \tau_1)).$$

Using Eq. (3.16), the MLE of θ for fixed values of α and β is given by

$$\hat{\theta}(\alpha, \beta) = \left(\frac{W}{r}\right)^{1/\alpha}. \tag{3.17}$$

Now inserting $\hat{\theta}(\alpha, \beta)$ instead of θ in Eqs. (3.14), (3.15) one can easily see that the MLEs of α and β can be obtained by solving the following two nonlinear equations simultaneously:

$$\frac{r - n_1}{\beta} + (\alpha - 1) \sum_{i=n_1+1}^{r} \frac{t_{i:n} - \tau_1}{\tau_1 + \beta(t_{i:n} - \tau_1)} - r\frac{W_1(\alpha, \beta)}{W(\alpha, \beta)} = 0,$$

$$\frac{r}{\alpha} + \sum_{i=1}^{n_1} \ln t_{i:n} + \sum_{i=n_1+1}^{r} \ln(\tau_1 + \beta(t_{i:n} - \tau_1)) - r\frac{W_2(\alpha, \beta)}{W(\alpha, \beta)} = 0.$$

Then $\hat{\alpha}$ and $\hat{\beta}$, the MLEs of α and β, respectively, can be substituted in Eq. (3.17) to obtain the MLE of θ. Initial values for the iterative procedure to solve the previous nonlinear equations can be found using the same technique as that of the Type-I censoring scheme.

Confidence intervals

As the MLEs do not exist in closed form, it is very difficult to obtain their exact distributions. Therefore, Abdel-Ghaly [109] suggested using the asymptotic confidence intervals based on the normal approximation. This procedure requires the inverse of the observed Fisher information matrix. The observed Fisher information matrix of (α, β, θ) in the presence of a Type-I censoring scheme is given by

$$I_{obs}(\alpha, \beta, \theta) = \begin{bmatrix} O_{11}^{(I)} & O_{12}^{(I)} & O_{13}^{(I)} \\ O_{21}^{(I)} & O_{22}^{(I)} & O_{23}^{(I)} \\ O_{31}^{(I)} & O_{32}^{(I)} & O_{33}^{(I)} \end{bmatrix}, \tag{3.18}$$

where

$$O_{11}^{(I)} = -\frac{n_1 + n_2}{\alpha^2} + \frac{\ddot{Q}_{\alpha\alpha} - 2\dot{Q}_\alpha \ln \theta + Q(\ln \theta)^2}{\theta^\alpha},$$

$$O_{12}^{(I)} = \sum_{i=n_1+1}^{n_1+n_2} \frac{t_{i:n} - \tau_1}{\tau_1 \beta(t_{i:n} - \tau_1)} - \frac{\ddot{Q}_{\alpha\beta} - \dot{Q}_\beta \ln \theta}{\theta^\alpha},$$

$$O_{13}^{(I)} = -\frac{n_1 + n_2}{\theta} + \frac{Q - \alpha Q \ln \theta}{\theta^{\alpha+1}},$$

$$O_{21}^{(I)} = \sum_{n_1+1}^{n_1+n_2} \frac{t_{i:n} - \tau_1}{\tau_1 + \beta(t_{i:n} - \tau_1)} - \frac{\ddot{Q}_{\beta\alpha} - \dot{Q}_{\beta}\ln\theta}{\theta^\alpha},$$

$$O_{22}^{(I)} = -\frac{n_2}{\beta^2} - (\alpha - 1) \sum_{i=n_1+1}^{n_1+n_2} \left(\frac{t_{i:n} - \tau_1}{\tau_1 + \beta(t_{i:n} - \tau_1)}\right)^2 - \frac{\ddot{Q}_{\beta\beta}}{\theta^\alpha},$$

$$O_{23}^{(I)} = -\frac{\alpha\dot{Q}_{\beta}}{\theta^{\alpha+1}},$$

$$O_{31}^{(I)} = -\frac{n_1 + n_2}{\theta} + \frac{\theta Q + \alpha\theta\dot{Q}_{\alpha} - \alpha(\alpha + 1)Q}{\theta^{\alpha+2}},$$

$$O_{32}^{(I)} = \frac{\alpha\dot{Q}_{\beta}}{\theta^{\alpha+1}},$$

$$O_{33}^{(I)} = \frac{\alpha(n_1 + n_2)}{\theta^2} - \frac{\alpha(\alpha + 1)Q}{\theta^{\alpha+2}},$$

$$\ddot{Q}_{\alpha\alpha} = \sum_{i=1}^{n_1} t_{i:n}^\alpha (\ln t_{i:n})^2 + \sum_{i=n_1+1}^{n_1+n_2} (\tau_1 + \beta(t_{i:n} - \tau_1))^\alpha (\ln(\tau_1 + \beta(t_{i:n} - \tau_1)))^2$$

$$+ (n - n_1 - n_2)(\tau_1 + \beta(\eta - \tau_1))^\alpha (\ln(\tau_1 + \beta(\eta - \tau_1)))^2,$$

$$\ddot{Q}_{\alpha\beta} = \ddot{Q}_{\beta\alpha} = \frac{\dot{Q}_{\beta}}{\alpha} + \alpha \sum_{i=n_1+1}^{n_1+n_2} (t_{i:n} - \tau_1)(\tau_1 + \beta(t_{i:n} - \tau_1))^{\alpha-1} \ln(\tau_1 + \beta(t_{i:n} - \tau_1))$$

$$+ \alpha(n - n_1 - n_2)(\eta - \tau_1)(\tau_1 + \beta(\eta - \tau_1))^{\alpha-1} \ln(\tau_1 + \beta(\eta - \tau_1)),$$

$$\ddot{Q}_{\beta\beta} = \alpha(\alpha - 1)\left[\sum_{i=n_1+1}^{n_1+n_2} (t_{i:n} - \tau_1)^2 (\tau_1 + \beta(t_{i:n} - \tau_1))^2\right.$$

$$\left. + (\eta - \tau_1)^2 (\tau_1 + \beta(\eta - \tau_1))^2 \right].$$

The asymptotic two-sided $100(1 - \gamma)\%$ confidence intervals of α, β, and θ can be obtained as

$$\hat{\alpha} \pm z_{\gamma/2}\sqrt{V(\hat{\alpha})}, \quad \hat{\beta} \pm z_{\gamma/2}\sqrt{V(\hat{\beta})}, \quad \text{and} \quad \hat{\theta} \pm z_{\gamma/2}\sqrt{V(\hat{\theta})},$$

respectively, where $V(\hat{\alpha})$, $V(\hat{\beta})$, and $V(\hat{\theta})$ are the diagonal entries of the inverse of I_{obs} as defined in Eq. (3.18). Similarly, using the likelihood function in Eq. (3.13), the observed Fisher information matrix can be obtained, which may be used to construct the asymptotic confidence intervals when the data are Type-II censored.

Abdel-Ghaly et al. [109] performed an extensive numerical study to judge the performance of MLEs of the parameters and the associated asymptotic confidence intervals. It has been noticed that the performance of MLEs and confidence intervals are quite satisfactory. The biases and mean squared errors (MSEs) of the MLEs of all the parameters decrease as the sample size increases. Also, the average length of the asymptotic confidence interval decreases as the sample size increases. However, the authors do not report the coverage percentages of the confidence intervals, though they provide meaningful information about the performance of the confidence interval.

3.2.3 Other distributions

In this section, we will briefly discuss the available literature that is based on the other distributional assumptions on T. We will mainly focus on the generalized exponential, generalized Weibull, Pareto Type-I, Lomax, and log-normal distributions.

Generalized exponential distribution

Analysis of simple SSLT data in the presence of Type-II censoring was considered by Ismail [111], when the lifetime at the initial stress level is assumed to have a generalized exponential distribution. The PDF of the lifetime at the initial stress level is given by

$$f_T(t) = \begin{cases} \alpha \lambda e^{-\lambda t} \left(1 - e^{-\lambda t}\right)^{\alpha-1} & \text{if } t > 0 \\ 0 & \text{otherwise,} \end{cases}$$

where $\alpha > 0$ and $\lambda > 0$ are called shape and scale parameters, respectively. The PDF of $\widetilde{T}$ is given by

$$f_{\widetilde{T}}(t) = \begin{cases} \alpha \lambda e^{-\lambda t} \left(1 - e^{-\lambda t}\right)^{\alpha-1} & \text{if } 0 < t \leq \tau_1 \\ \beta \alpha \lambda e^{-\lambda(\tau_1 + \beta(t-\tau_1))} \left(1 - e^{\lambda(\tau_1 + \beta(t-\tau_1))}\right)^{\alpha-1} & \text{if } \tau_1 < t < \infty \\ 0 & \text{otherwise.} \end{cases}$$

Given Type-II censored data, the log likelihood function, ignoring the additive constant term, can be expressed as

$$l(\alpha, \lambda, \beta) = r(\ln \alpha + \ln \lambda) + (r - n_1) \ln \beta - \lambda \left(\sum_{i=1}^{n_1} t_{i:n} + \sum_{i=n_1+1}^{r} (\tau_1 + \beta(t_{i:n} - \tau_1)) \right)$$

$$+ (\alpha - 1) \left(\sum_{i=1}^{n_1} \ln \left(1 - e^{-\lambda t_{i:n}}\right) + \sum_{i=n_1+1}^{r} \ln \left(1 - e^{-\lambda(\tau_1 + \beta(t_{i:n} - \tau_1))}\right) \right)$$

$$+ (n - r) \ln \left(1 - \left(1 - e^{-\lambda(\tau_1 + \beta(t_{r:n} - \tau_1))}\right)^{\alpha}\right).$$

The MLEs of unknown parameters can be found by maximizing the log likelihood function with respect to α, λ, and β.

Generalized Weibull distribution

The CDF and PDF of a generalized Weibull distribution with the parameters $\alpha > 0$, $\gamma > 0$ and $\theta > 0$ are given by

$$F(t) = \begin{cases} 1 - e^{1-(1+(t/\theta)^{\alpha})^{\gamma}} & \text{if } t \geq 0 \\ 0 & \text{otherwise,} \end{cases}$$

and

$$f(t) = \begin{cases} \frac{\alpha\gamma}{\theta^{\alpha}} t^{\alpha-1} \left(1 + (t/\theta)^{\alpha}\right)^{\gamma-1} e^{1-(1+(t/\theta)^{\alpha})^{\gamma}} & \text{if } t \geq 0 \\ 0 & \text{otherwise,} \end{cases}$$

respectively. This distribution was first proposed by Bagdonavicius and Nikulin [2] to model failure time data. All the moments and the moment generating function (MGF) of a generalized Weibull distribution exist for all values of $\alpha > 0$, $\gamma > 0$, and $\theta > 0$. The main advantage of the generalized Weibull distribution is the flexibility of the model to fit a large verity of lifetime data. The hazard function corresponding to the distribution can be expressed as

$$h(t) = \frac{\alpha}{\gamma\theta^{\alpha}} t^{\alpha-1} \left(1 + (t/\theta)^{\alpha}\right)^{\gamma-1} \quad \text{for } t > 0,$$

which can take various shapes including constant, monotone, reverse-U shape, and bathtub shape. For example, the shapes of the $h(t)$ for different values of the parameters are given in the following table.

Parameter range	Shape of $h(t)$
$\alpha > \max\{1, 1/\gamma\}$	Increases from zero to ∞
$\alpha = 1$ and $\gamma > 1$	Increases from γ/θ to ∞
$0 < \alpha < \max\{1, 1/\gamma\}$	Decreases from ∞ to zero
$0 < \alpha < 1$ and $\gamma = 1/\alpha$	Decreases from ∞ to $1/\theta$

Recently, the analysis of a simple SSLT in the presence of a progressive Type-II censoring scheme has been addressed by El-Din et al. [112]. The authors assumed that the distribution of T is a generalized Weibull distribution with $\theta = 1$. One can find the PDF of $\widetilde{T}$ using Eq. (1.4). Given a progressive Type-II censoring scheme $R_1, \ldots, R_m$ and a simple SSLT data $t_{1:n} < \cdots < t_{n_1:n} < \tau_1 < t_{n_1+1:n} < \cdots < t_{m:n}$ under that scheme the log-likelihood function, ignoring the additive constant term, of the model parameters can be written as

$$l(\alpha, \gamma, \beta) = m\ln(\alpha\gamma) + (m - n_1)\ln\beta$$

$$+ (\alpha - 1)\left(\sum_{i=1}^{n_1}\ln t_{i:n} + \sum_{i=n_1+1}^{m}\ln\psi_i(\beta)\right)$$

$$+ (\gamma - 1) \left(\sum_{i=1}^{n_1} \ln(1 + t_{i:n}^\alpha) + \sum_{i=n_1+1}^{m} \ln\left(1 + \psi_i^\alpha(\beta)\right) \right)$$

$$+ \sum_{i=1}^{n_1} (R_i + 1)\left(1 - (1 + t_{i:n}^\alpha)^\gamma\right) + \sum_{i=n_1+1}^{m} (R_i + 1)\left(1 - (1 + \psi_i^\alpha(\beta))^\gamma\right),$$

where $\psi_i(\beta) = \tau_1 + \beta(t_{i:n} - \tau_1)$. Taking partial derivatives of the log-likelihood function with respect to α, γ, and β, and equating them to zero, one can obtain the normal equations. The MLEs of the unknown model parameters can be found by solving the system of normal equations. The asymptotic confidence intervals can be obtained using the diagonal entries of the inverse of the observed Fisher information matrix. Alternatively, a bootstrap confidence intervals may be constructed in a routine manner.

El-Din et al. [112] also considered the inferential issues under the Bayesian set-up. The authors assumed two sets of prior distributions. The first prior can be described as follows. The parameters α, γ, and β are assumed to be independent a priori with prior PDFs

$$\pi_\alpha^{\text{NIP}}(\alpha) = \alpha^{-1} \quad \text{for } \alpha > 0,$$
$$\pi_\gamma^{\text{NIP}}(\gamma) = \gamma^{-1} \quad \text{for } \gamma > 0,$$
$$\pi_\beta^{\text{NIP}}(\beta) = \beta^{-1} \quad \text{for } \beta > 1,$$

and hence, the joint prior PDF is given by

$$\pi_{\text{NIP}}(\alpha, \gamma, \beta) = (\alpha\gamma\beta)^{-1} \quad \text{for } \alpha > 0, \gamma > 0, \beta > 1.$$

Note that it is a noninformative and improper prior. Now we will describe the second prior which was considered by the authors. The parameter α has a one-parameter exponential distribution with mean λ, and for given α, the conditional distribution of γ is a Gamma$(\mu, 1/\alpha)$. The parameter β is assumed to be independent of α and γ and β has a prior density $\pi_\beta^{\text{NIP}}(\cdot)$. The joint prior PDF can be expressed as

$$\pi_{\text{IP}}(\alpha, \gamma, \beta) \propto \beta^{-1} \alpha^{-\mu} \gamma^{\mu-1} e^{-\alpha/\lambda - \gamma/\alpha} \quad \text{for } \alpha > 0, \gamma > 0, \beta > 1.$$

Note that this is an informative prior, but it is improper. The joint posterior PDF can be obtained by multiplying the joint prior PDF and the likelihood function. The joint posterior PDFs are given by

$$\pi_{\text{NIP}}^*(\alpha, \gamma, \beta | \text{Data}) \propto (\alpha\gamma)^{m-1} \beta^{m-n_1-1}$$

$$\times \left\{ \prod_{i=1}^{n_1} t_{i:n}^{\alpha-1} (1 + t_{i:n}^\alpha)^{\gamma-1} e^{(R_i+1)(1-(1+t_{i:n}^\alpha)^\gamma)} \right\}$$

$$\times \left\{ \prod_{i=n_1+1}^{m} \psi_i^{\gamma-1}(\beta)(1 + \psi_i^\alpha(\beta))^{\gamma-1} e^{(R_i+1)(1-(1+\psi_i^\alpha(\beta))^\gamma)} \right\}$$

and

$$\pi_{\mathrm{IP}}^*(\alpha,\,\gamma,\,\beta|\mathrm{Data}) \propto \alpha^{m-\mu}\gamma^{m+\mu-1}\beta^{m-n_1-1}e^{-\alpha/\lambda-\gamma/\alpha}$$

$$\times \left\{ \prod_{i=1}^{n_1} t_{i:n}^{\alpha-1}(1+t_{i:n}^{\alpha})^{\gamma-1}e^{(R_i+1)\left(1-(1+t_{i:n}^{\alpha})^{\gamma}\right)} \right\}$$

$$\times \left\{ \prod_{i=n_1+1}^{m} \psi_i^{\gamma-1}(\beta)(1+\psi_i^{\alpha}(\beta))^{\gamma-1}e^{(R_i+1)\left(1-(1+\psi_i^{\alpha}(\beta))^{\gamma}\right)} \right\},$$

respectively. The authors proposed to obtain the Bayes estimation (BE) based on two methods, viz., Lindley's approximation and the Markov chain Monte Carlo (MCMC) technique. The full conditional distributions corresponding to the posterior distributions can be easily obtained and they can be used for the Metropolis technique. The authors suggested using the normal distribution as the proposal distribution.

El-din et al. [112] performed a simulation study to check the performance of the different estimators proposed in the article. It is observed that the MSE of the MLEs and BEs decreases as the effective sample size m increases. The BEs perform better than the MLEs with respect to their MSEs. For large values of m, the BEs obtained using the MCMC technique are better than the BEs obtained using the Lindley's approximation. For small values of m, the Lindley's approximation works better than the MCMC technique for the parameter β, whereas for the other two parameters the MCMC technique performs better than the Lindley's approximation. Among the classical interval estimation techniques, the performance of the bootstrap method is better than the asymptotic method for all the parameters as the average lengths are larger in the latter case. The authors did not report the coverage percentages of the intervals, which along with their avarage lengths provides valuable information for meaningful comparisons of the performance of the interval estimators. Also, the Bayesian credible interval (CRI) was not considered in the article. One can easily obtain the CRIs of different unknown parameters using the realization of the Markov chain having the posterior distribution as the equilibrium distribution. Given the fact that the BEs perform better than MLEs, it is expected that the Bayesian CRI might works better than the corresponding bootstrap confidence intervals.

Pareto Type-I distribution

The distribution of T is assumed to be a Pareto Type-I, i.e., the PDF of T is given by

$$f_T(t) = \frac{\alpha\theta^{\alpha}}{t^{\alpha+1}}\mathbb{1}_{[\theta,\infty)}(t),$$

where $\alpha > 0$ and $\theta > 0$. The classical inference of the TRVM on the assumption that T has a Pareto Type-I was discussed by Shi and Shi [113] in the presence of the progressively hybrid Type-I censoring scheme. Progressive Type-I censored data from a simple SSLT can have the following forms:

$$t_{1:n} < \cdots < t_{n_1:n} < \tau_1 < t_{n_1+1:n} < \cdots < t_{n_1+n_2:n},$$

where n_2 is the number of failures before τ_2 at the second stress level if $t_{m:n} > \tau_2$, and $n_2 = m - n_1$ if $t_{m:n} \leq \tau_2$. Note that n_1 or n_2 can be zero. Hence the log-likelihood function, ignoring the additive constant term, can be written as

$$l(\alpha, \theta, \beta) = (n_1 + n_2) \ln \alpha + n\alpha \ln \theta + n_2 \ln \beta - \sum_{i=1}^{n_1} (\alpha R_i + \alpha + 1) \ln t_{i:n}$$

$$- \sum_{i=n_1+1}^{n_1+n_2} (\alpha R_i + \alpha + 1) \ln (\tau_1 + \beta(t_{i:n} - \tau_1))$$

$$- \alpha R \ln (\tau_1 + \beta(M - \tau_1)),$$

where $R = n - m - \sum_{i=1}^{m-1} R_i$ if $t_{m:n} \leq \tau_2$, otherwise $R = n - n_1 - n_2 - \sum_{i=1}^{n_1+n_2} R_i$, and $M = \min\{t_{m:n}, \tau_2\}$. Note that for fixed values of θ and β, $l(\alpha, \theta, \beta)$ is an increasing function of α. Hence the MLE of α is $t_{1:n}$. Now, $p(\theta, \beta) = l(t_{1:n}, \theta, \beta)$ can be maximized to find the MLEs of θ and β. Though Shi and Shi [113] suggested constructing the asymptotic confidence intervals of the unknown parameters using the observed Fisher information matrix, it may be noted that the first order moment of the Pareto Type-I distribution does not exist and hence the regularity conditions of the asymptotic normality of MLE may not hold. One may construct parametric bootstrap confidence intervals of the unknown parameters in a simple and routine manner.

Lomax distribution

El-Sagheer and Ahsanullah [114] considered the simple SSLT under the assumption that T has a Lomax distribution, i.e., the PDF of T is given by

$$f_T(t) = \frac{\alpha \theta^\alpha}{(\theta + t)^{\alpha+1}} \mathbb{1}_{(0,\infty)}(t),$$

where $\alpha > 0$ and $\theta > 0$ are the shape and scale parameters, respectively. Note that the Lomax distribution is also known as the Pareto Type-II distribution. Given the progressive Type-II censored data, the log-likelihood function (ignoring the additive constant term) of the parameters $\alpha > 0$ and $\theta > 0$ can be expressed as

$$l(\alpha, \theta, \beta) = m \ln \alpha + \alpha \sum_{i=1}^{m} (R_i + 1) \ln \theta + (m - n_1) \ln \beta$$

$$- \sum_{i=1}^{n_1} (\alpha R_i + \alpha + 1) \ln(\theta + t_{i:n})$$

$$- \sum_{i=n_1+1}^{m} (\alpha R_i + \alpha + 1) \ln(\theta + \tau_1 + \beta(t_{i:n} - \tau_1)).$$

Taking partial derivatives of the log-likelihood function with respect to the unknown parameters α, θ, and β and equating them to zero, one can obtain the normal equations. The MLEs of the unknown model parameters can be found by solving the system of normal equations. El-Sagheer and Ahsanullah [114] considered asymptotic confidence intervals of the unknown parameters based on the observed Fisher information matrix. However, as the first order moment of a Lomax distribution does not exist, the necessary regularity conditions for consistency and asymptotic normality of the MLE may not hold true for this distribution. Therefore, a detailed and formal proof of the same would be more convincing. Bootstrap confidence intervals can be constructed using the algorithm provided by El-Sagheer and Ahsanullah [114].

The Bayesian inferential procedure of the same model was also constructed by El-Sagheer and Ahsanullah [114] under the assumption of noninformative priors on the parameters. It is also assumed that the parameters are independently distributed apriori. The joint posterior PDF of α, θ, β can be written as

$$\pi^*(\alpha, \theta, \beta | \text{Data}) \propto \frac{\alpha^{m-1} \theta^{\alpha \sum_{i=1}^{m}(R_i+1)-1} \beta^{m-n_1-1}}{\prod_{i=1}^{n_1}(\theta + t_{i:n})^{\alpha(R_i+1)-1} \prod_{i=n_1}^{m}(\theta + \tau_1 + \beta(t_{i:n} - \tau_1))^{\alpha(R_i+1)-1}},$$

for $\alpha > 0$, $\theta > 0$, and $\beta > 1$. It is noticed that the BE of a parametric function does not exist in explicit form. The authors exploited the full conditional distributions to compute the BE and construct a CRI of a function of the unknown parameters using the MCMC technique.

Log-normal distribution

The classical inferential issues of the TRVM under the assumption that T has a log-normal distribution with the PDF

$$f_T(t) = \frac{1}{\sigma t \sqrt{2\pi}} e^{-\frac{1}{2}\left(\frac{\ln t - \mu}{\sigma}\right)^2} \mathbb{1}_{(0, \infty)}(t),$$

where $\mu \in \mathbb{R}$ and $\sigma > 0$, was addressed by Bai et al. [115]. The authors also assumed that the data are Type-I censored. Under the log-normal distribution, the PDF of $\tilde{T}$ is given by

$$f_{\tilde{T}}(t) = \begin{cases} \frac{1}{\sigma t \sqrt{2\pi}} e^{-\frac{1}{2}\left(\frac{\ln t - \mu}{\sigma}\right)^2} & \text{if } 0 < t \leq \tau_1 \\ \frac{\beta}{\sigma(\tau_1 + \beta(t - \tau_1))\sqrt{2\pi}} e^{-\frac{1}{2}\left(\frac{\ln(\tau_1 + \beta(t - \tau_1)) - \mu}{\sigma}\right)^2} & \text{if } t > \tau_1 \\ 0 & \text{otherwise.} \end{cases}$$

Given a Type-I censored data $t_{1:n} < \cdots < t_{n_1:n} < \tau_1 < t_{n_1+1:n} < \cdots < t_{n_1+n_2:n} < \eta$, the log-likelihood function can be written as

$$l(\mu, \sigma^2, \beta) = -\frac{n_1 + n_2}{2} \ln \sigma^2 - \sum_{i=n_1+1}^{n_1+n_2} \ln \left(\tau_1 + \beta(t_{i:n} - \tau_1)\right)$$

$$- \frac{1}{2\sigma^2} \sum_{i=1}^{n_1} (\ln t_{i:n} - \mu)^2 - \frac{1}{2\sigma^2} \sum_{i=n_1+1}^{n_1+n_2} (\ln(\tau_1 + \beta(t_{i:n} - \tau_1)) - \mu)^2$$

$$+ (n - n_1 - n_2) \ln \left(1 - \Phi\left(\frac{\ln(\tau_1 + \beta(\eta - \tau_1)) - \mu}{\sigma}\right)\right),$$

where $\Phi(\cdot)$ is the CDF of the standard normal distribution. The MLEs of μ, σ^2, and β can be found by maximizing the log-likelihood function with respect to the parameters. Note that in this case the MLEs of the unknown parameters do not exist in explicit form and one needs to use a numerical technique to maximize the log-likelihood function. The asymptotic variance of the MLEs and the confidence intervals of the unknown parameters can be obtained using the observed Fisher information matrix.

3.3 Tempered failure rate model

As discussed in Section 1.5.3, the TFRM assumes that the effect of changing the stress level from s_1 to s_2 at time τ_1 is equivalent to multiplying the failure rate function (FRF) at the stress level s_1 by an unknown positive constant β. Suppose that $\lambda(\cdot)$ and $\tilde{\lambda}(\cdot)$ are the failure rates at the first stress level s_1 and under the whole step-stress pattern, respectively. Under the TFRM assumption, $\lambda(\cdot)$ and $\tilde{\lambda}(\cdot)$ satisfy the following relationship:

$$\tilde{\lambda}(t) = \begin{cases} \lambda(t) & \text{if } t \le \tau_1 \\ \beta\lambda(t) & \text{if } t > \tau_1. \end{cases}$$

Weibull distribution

Bhattacharyya and Soejoeti [21] considered the classical inference of the model parameters under the TFRM, when the distribution of the lifetime at the first stress level is assumed to have a Weibull distribution with the shape parameter $\alpha > 0$ and scale parameter $\theta > 0$ having the following PDF:

$$f(t) = \frac{\alpha}{\theta^\alpha} t^{\alpha-1} e^{-(t/\theta)^\alpha} \mathbb{1}_{[0,\infty)}(t). \tag{3.19}$$

The corresponding failure rate function (FRF) is given by

$$\lambda(t) = \frac{\alpha}{\theta^\alpha} t^{\alpha-1} \quad \text{for } t \ge 0.$$

Hence using Eqs. (1.5), (1.6), the FRF and the PDF of the lifetime under the whole step-stress pattern are given by

$$\tilde{\lambda}(t) = \frac{\alpha}{\theta^\alpha} t^{\alpha-1} \left(\mathbb{1}_{[0,\tau_1]}(t) + \beta\mathbb{1}_{(\tau_1,\infty)}(t)\right),$$

and

$$
\tilde{f}(t) = \begin{cases} \frac{\alpha}{\theta^\alpha} t^{\alpha-1} e^{-(t/\theta)^\alpha} & \text{if } 0 \le t \le \tau_1 \\ \frac{\beta\alpha}{\theta^\alpha} t^{\alpha-1} e^{-\beta(t/\theta)^\alpha - (1-\beta)(\tau_1/\theta)^\alpha} & \text{if } t > \tau_1 \\ 0 & \text{otherwise,} \end{cases}
$$

respectively. Given a complete sample $\{t_{1:n} < \cdots < t_{n_1:n} < \tau_1 < t_{n_1+1:n} < \cdots < t_{n:n}\}$, the log-likelihood function of α, θ, and β, ignoring the additive constant term, can be written as

$$
l(\alpha, \theta, \beta) = n \ln \alpha + (n - n_1) \ln \beta - n\alpha \ln \theta + \alpha \sum_{i=1}^{n} \ln t_{i:n} - \frac{1}{\theta^\alpha} (A(\alpha) + \beta B(\alpha)),
$$

where $A(\alpha) = \sum_{i=1}^{n_1} t_{i:n}^\alpha + (n - n_1)\tau_1^\alpha$ and $B(\alpha) = \sum_{i=n_1+1}^{n} t_{i:n}^\alpha - (n - n_1)\tau_1^\alpha$. It is quite easy to show that the MLEs of α, θ, and β do not exist if $n_1 = 0$ or $n_1 = n$. The normal equations can be obtained by taking the partial derivatives of the log-likelihood function with respect to α, θ, and β and equating them to zero. For $0 < n_1 < n$, the normal equations are given by

$$
\dot{l}_\alpha = \frac{n}{\alpha} - n \ln \theta - \frac{1}{\theta^\alpha} (\dot{A}_\alpha + \beta \dot{B}_\alpha) = 0, \tag{3.20}
$$

$$
\dot{l}_\theta = \frac{n\alpha}{\theta} - \frac{\alpha}{\theta^{\alpha+1}} (A(\alpha) + \beta B(\alpha)) = 0, \tag{3.21}
$$

$$
\dot{l}_\beta = \frac{n - n_1}{\beta} - \frac{B(\alpha)}{\theta^\alpha} = 0, \tag{3.22}
$$

where

$$
\dot{A}_\alpha = \sum_{i=1}^{n_1} t_{i:n}^\alpha \ln t_{i:n} + (n - n_1)\tau_1^\alpha \ln \tau_1,
$$

$$
\dot{B}_\alpha = \sum_{i=n_1+1}^{n} t_{i:n}^\alpha \ln t_{i:n} - (n - n_1)\tau_1^\alpha \ln \tau_1.
$$

For a fixed α, the MLEs of θ and β can be found explicitly by solving Eqs. (3.21), (3.22) simultaneously. They are given by

$$
\hat{\theta}(\alpha) = \left(\frac{A(\alpha)}{n_1}\right)^{1/\alpha} \quad \text{and} \quad \hat{\beta}(\alpha) = \frac{n - n_1}{n} \times \frac{A(\alpha)}{B(\alpha)}. \tag{3.23}
$$

Substituting $\hat{\theta}(\alpha)$ and $\hat{\beta}(\alpha)$ for θ and β, respectively, in Eq. (3.20) yields

$$
\frac{n_1}{n} \times \frac{\dot{A}_\alpha}{A(\alpha)} + \frac{n - n_1}{n} \times \frac{\dot{B}_\alpha}{B(\alpha)} - \frac{1}{\alpha} = \frac{1}{n} \sum_{i=1}^{n} \ln t_{i:n}. \tag{3.24}
$$

One needs to use an iterative numerical method, like the Newton-Raphson method, to solve Eq. (3.24). The solution of Eq. (3.24) provides the MLE of α. Once the MLE of α is obtained, it can be substituted in Eq. (3.23) to get the MLEs of θ and β. The initial value of α for the iterative methods can be found using the following logic. The natural starting point is to use a one-parameter exponential distribution and the initial value of α may be considered as one. Alternatively, one can fit a linear regression model $y_i = v_0 + v_1 x_i, i = 1, 2 \ldots, n_1$, where $x_i = \ln t_{i:n}$ and $y_i = \ln(\ln n - \ln(n - i + 0.5))$. Let $\hat{v}_0$ be the least squares estimate of v_0. Then $e^{\hat{v}_0}$ may be taken as the initial value of α for the iterative method to solve Eq. (3.24).

Madi [23] generalized the TFRM by incorporating multiple step-stress levels by assuming that the FRF, $\tilde{\lambda}(t)$, under the multiple SSLT is given by

$$\tilde{\lambda}(t) = \left(\prod_{j=0}^{i-1} \beta_j \right) \lambda(t) \quad \text{if} \quad \tau_{i-1} \le t < \tau_i \quad i = 1, \ldots, k,$$

where $\tau_0 = 0$, $\tau_k = \infty$, $\beta_0 = 1$, and $\beta_i > 0$ for $i = 1, 2, \ldots, k - 1$. Assuming that the lifetime at the first stress level has a Weibull distribution having the PDF as given in Eq. (3.19), Madi [23] obtained the MLEs of α, θ and β_i, $i = 1, 2, \ldots, k - 1$ based on a complete sample. He showed that the problem of finding the MLEs can be reduced to solving a single nonlinear equation involving α. It can be shown that the MLEs under this set-up are unique; see Wang and Fei [116] for the proof. Wang and Fei [116] conducted a simulation study to judge the accuracy of the MLEs in the presence of a simple SSLT. It can be seen that the biases and standard errors of the MLEs of the parameters are quite small and the performance of the MLEs are quite satisfactory.

Khamis and Higgins [24] considered the inferential issues of the same model as assumed by Madi [23]. To describe the relationship between the stress levels and the unknown scale parameters, the authors assumed the log-linear link function, i.e.,

$$\ln \theta_i = v_0 + v_1 s_i, \quad \text{where} \quad \theta_i = \theta \prod_{j=0}^{i-1} \frac{1}{\beta_j}, \quad i = 1, 2, \ldots, k.$$

The authors also assumed that the data are Type-I censored at the prefixed time η. Ignoring the additive constant term, the log-likelihood function of α, v_0, and v_1 is given by

$$l(\alpha, v_0, v_1) = \tilde{n}_k \ln \alpha - \tilde{n}_k v_0 - v_1 \sum_{i=1}^{k} n_i s_i + \alpha \sum_{i=1}^{\tilde{n}_k} \ln t_{i:n} - \sum_{i=1}^{k} A_i(\alpha) e^{-(v_0 + v_1 s_i)},$$

where

$$A_i(\alpha) = \sum_{j=\tilde{n}_{i-1}+1}^{\tilde{n}_i} \left(t_{i:n}^{\alpha} - \tau_{i-1}^{\alpha} \right) + (n - \tilde{n}_i) \left(\tau_i^{\alpha} - \tau_{i-1}^{\alpha} \right) \quad \text{and} \quad \tilde{n}_i = \sum_{j=1}^{i} n_j.$$

One can obtain the normal equations by taking the partial derivatives of the log-likelihood function and equating them to zero. A numerical technique needs to be used to solve the system of three nonlinear equations simultaneously.

Sha and Pan [25] considered the Bayesian estimation of the parameters based on the data form a SSLT in presence of PCS-I. It is assumed that the test is started at the stress level s_1. Let R_i, $i = 1, 2, \ldots, k - 1$, be prefixed integers, and R_i items are randomly chosen from the survival items at the time τ_i and removed from the test at the time τ_i. At the time η ($> \tau_{k-1}$) all the survived items are removed from the test and the test is terminated. The analysis is performed under this censoring scheme based on the assumption that the scheme is feasible; see Balakrishnan [12]. Sha and Pan [25] assumed that the distribution of the lifetime under the SSLT is given by the Khamis-Higgins model (see Khamis and Higgins [117]), which is a special case of the TFRM. The authors also assumed the log-linear link function between the scale parameters and stress levels, i.e., $\ln(\theta_i) = v_0 + v_1 s_i$. The likelihood function of parameters for $\alpha > 0$, $v_0 \in \mathbb{R}$, and $v_1 \in \mathbb{R}$ is given by

$$L(\alpha, v_0, v_1) = \alpha^{\widetilde{n}_k} \left(\prod_{i=1}^{\widetilde{n}_k} t_{i:n} \right) e^{L_1(\alpha, v_0, v_1)},$$

where

$$L_1(\alpha, v_0, v_1) = \widetilde{n}_k v_0 + v_1 \sum_{i=1}^{k} n_i s_i - \sum_{i=1}^{k} e^{v_0 + v_1 s_i} \widetilde{A}_i(\alpha),$$

$$\widetilde{A}_i(\alpha) = \sum_{j=\widetilde{n}_{i-1}+1}^{\widetilde{n}_i} \left(t_{i:n}^{\alpha} - \tau_{i-1}^{\alpha} \right) + \left(n - \widetilde{n}_i - \widetilde{R}_{i-1} \right) \left(\tau_i^{\alpha} - \tau_{i-1}^{\alpha} \right),$$

$$\widetilde{n}_i = \sum_{j=1}^{i} n_j, \quad \text{and} \quad \widetilde{R}_i = \sum_{j=1}^{i} R_j, \quad i = 1, 2, \ldots, k.$$

The authors adopted independent priors for the parameters α, v_0, and v_1. It is noticed that the conjugate prior on v_0 is a generalized extreme value distribution having PDF

$$\pi_1(v_0) \propto e^{\kappa_0 v_0 - \gamma_0 e^{v_0}} \mathbb{1}_{(0, \infty)}(v_0),$$

where $\kappa_0 > 0$ and $\gamma_0 > 0$ are the hyperparameters. Note that κ_0 and γ_0 are the shape and scale parameters, respectively. Conjugate priors do not exist for v_1 and α. However, the authors consider the conditional likelihood function of v_1 when v_0 and α are known and chose a prior on v_1 whose PDF has similar functional form to its conditional likelihood function. A similar argument is used to choose the prior on α. The prior PDFs of v_1 and α are given by

$$\pi_2(v_1) \propto e^{\kappa_1 v_1 - \gamma_1 e^{\delta_1 v_1}} \mathbb{1}_{(0, \infty)}(v_1) \quad \text{and} \quad \pi_3(\alpha) \propto e^{\kappa_2 \alpha - \gamma_2 e^{\delta_2 \alpha}} \mathbb{1}_{(0, \infty)}(\alpha),$$

respectively, where the hyperparameters κ_1, κ_2, γ_1, γ_2, δ_1, and δ_2 are positive real numbers. It is also assumed that ν_0, ν_1, and α are independently distributed a priori. The full conditional posterior PDFs are given by

$$f_1(\nu_0|\nu_1, \alpha, \text{Data}) \propto \exp\left[(\kappa_0 + \tilde{n}_k)\, \nu_0 - \left(\gamma_0 + \sum_{i=1}^{k} e^{\nu_1 s_i} A_i(\alpha) \right) e^{\nu_0} \right],$$

$$f_2(\nu_1|\nu_0, \alpha, \text{Data}) \propto \exp\left[\left(\kappa_1 + \sum_{i=1}^{k} n_i s_i \right) \nu_1 - \left(\gamma_1 e^{\delta_1 \nu_1} + \sum_{i=1}^{k} e^{\nu_0 + \nu_1 s_i} A_i(\alpha) \right) \right],$$

$$f_3(\alpha|\nu_0, \nu_1, \text{Data}) \propto \alpha^{\tilde{n}_k} \exp\left[\left(\kappa_2 + \sum_{i=1}^{\tilde{n}_k} \ln t_{i:n} \right) \alpha - \left(\gamma_2 e^{\delta_2 \alpha} + \sum_{i=1}^{k} e^{\nu_0 + \nu_1 s_i} A_i(\alpha) \right) \right],$$

respectively, where $\nu_0 \in \mathbb{R}$, $\nu_1 \in \mathbb{R}$, and $\alpha > 0$. The authors have suggested performing inference by drawing posterior samples through an MCMC technique using the Gibbs sampling algorithm. The authors have performed an extensive numerical study to judge the performance of the procedure described here.

3.4 Cumulative risk model

The CEM, TRVM, and TFRM do not guarantee the continuity of the hazard function at the points where the stress levels are changed. In fact for most of the distributions that are used in reliability and survival analysis, the hazard functions are discontinuous at the stress changing points under these models. A jump in the hazard function implies that the effect of the change of the stress level is instantaneous, which may not be very reasonable in practice. A more resonable model would be to assume that the effect of the change in the stress level gradually changes the failure rate of the item under consideration. For example, suppose that the stress level involves temperature in an experiment; then the environment of the experiment will take some time to reach the desired temperature. Drop et al. [27] consider a more realistic model to overcome this problem. Latter Kannan et al. [28] named the model as the CRM. The model assumes that the effect of increase in the stress level is not noticed immediately, but there is a latency period before the effects are completely observed. Let ζ_i be the latency period to notice the complete effects of the stress level s_{i+1} after τ_i. As discussed in Section 1.5.4, it is assumed that the hazard function of the lifetime under the step-stress pattern is piecewise continuous, whereas an increasing continuous function joining the hazard functions at τ_i and $\tau_i + \zeta_i$ is used to model the hazard function during the latency period ζ_i. Thus the hazard function of the lifetime under the whole step-stress pattern is given by

$$h(t) = \begin{cases} h_i(t) & \text{if } \tau_{i-1} + \zeta_{i-1} < t \le \tau_i \\ \rho_i(t) & \text{if } \tau_i < t \le \tau_i + \zeta_i, \end{cases}$$

where $i = 1, 2, \ldots, k$, $\tau_0 = \zeta_0 = 0$, and $\tau_k = \infty$. Here $\rho_i(\cdot)$ is an increasing and continuous function over $(\tau_i, \tau_i + \zeta_i]$ satisfying $\rho_i(\tau_i) = h_i(\tau_i)$ and $\rho_i(\tau_i + \zeta_i) = h_{i+1}(\tau_i + \zeta_i)$ for all $i = 1, 2, \ldots, k - 1$. The function $\rho_i(\cdot)$ is called the ramping function. Drop et al. [27], Drop and Mazzuchi [118], Kannan et al. [28], and Beltrami [119, 120] considered this model and addressed inferential issues under the assumption that ζ_i are known. All the authors assume that $\rho_i(\cdot)$ is a linear function of time. It is worth mentioning here that the assumption of an instantaneous jump in the hazard function, though unrealistic, leads to a more simple and tractable model.

Drop et al. [27] and Drop and Mazzuchi [118] considered a multiple step-stress life test, where the hazard function at each stress level is constant and addressed the Bayesian estimation of the unknown parameters. The hazard function under the step-stress pattern is given by

$$
h(t) = \begin{cases} \lambda_i & \text{if } \tau_{i-1} + \zeta_{i-1} < t \le \tau_i \\ \lambda_i + \frac{\lambda_{i+1} - \lambda_i}{\zeta_i} (t - \tau_i) & \text{if } \tau_i < t \le \tau_i + \zeta_i, \end{cases} \tag{3.25}
$$

where $i = 1, 2, \ldots, k$, $\tau_0 = \zeta_0 = 0$, and $\tau_k = \infty$. As the aim of an SSLT is to get quick failures, it is not very unrealistic to assume that the failure rate at the stress level s_i is smaller than that at the stress level s_{i+1}, which imposes the order restriction $0 < \lambda_1 < \lambda_2 < \cdots < \lambda_k$ on the parameters. The authors incorporate this information into the prior. Instead of assuming the prior PDF defined on the region $\lambda_1 < \lambda_2 < \cdots < \lambda_k$, the authors reparameterized the problem using the one-to-one transformation $\upsilon_i = e^{-c\lambda_i}$ where $c > 0$ is a prespecified constant. Then $1 > \upsilon_1 > \upsilon_2 > \cdots > \upsilon_k > 0$. A prior distribution which is mathematically tractable, defined on this region, and imposes no other restriction on υ_k's, is a multivariate ordered Dirichlet distribution. The prior PDF of $\boldsymbol{\upsilon} = (\upsilon_1, \upsilon_2, \ldots, \upsilon_k)$ is given by

$$
\pi(\boldsymbol{\upsilon}) = \begin{cases} \Gamma(\alpha) \prod_{i=1}^{k+1} \frac{(\upsilon_{i-1} - \upsilon_i)^{\alpha\beta_i - 1}}{\Gamma(\alpha\beta_i)} & \text{if } 0 = \upsilon_{k+1} < \upsilon_k < \cdots < \upsilon_1 < \upsilon_0 = 1 \\ 0 & \text{otherwise,} \end{cases}
$$

$$\tag{3.26}$$

where $\alpha > 0$ and $\beta_i > 0$ for all $i = 1, 2, \ldots, k + 1$ and $\sum_{i=1}^{k+1} \beta_i = 1$. It is worth mentioning here that the marginal prior distribution of υ_i is $\text{Beta}\left(\alpha(1 - \tilde{\beta}_i), \alpha\tilde{\beta}_i\right)$, where $\tilde{\beta}_i = \sum_{j=1}^{i} \beta_j$.

Let us consider a SSLT where the failures of the test item can be determined only at τ_i, $i = 1, 2, \ldots, \tau_{k-1}$. All the failed items are removed from the test and the test continues with the surviving items. Let n_i be the number of failure that occur between τ_{i-1} and τ_i. Let m_i be the number of surviving items at the time point τ_i. Then ignoring the multiplicative constant term, the likelihood function of $\boldsymbol{\upsilon}$ is given by

$$
L(\boldsymbol{\upsilon}) = \prod_{i=1}^{k} (\psi_i)^{m_i} (1 - \psi_i)^{n_i}, \tag{3.27}
$$

where

$$\psi_i = P\left(T > \tau_i | T > \tau_{i-1}\right) = \exp\left[-\int_{\tau_{i-1}}^{\tau_i} h(t)dt\right]$$

$$= (\upsilon_i)^{(2\tau_i - 2\tau_{i-1} - \zeta_{i-1})/c} \, (\upsilon_{i-1})^{\zeta_{i-1}/2c}.$$

Now one can write the posterior PDF of υ by taking the product of the prior PDF given in Eq. (3.26) and the likelihood function given in Eq. (3.27). The marginal posterior PDFs do not exist in a closed form. Also in most of the cases the Bayes estimate of some parametric function do not have an explicit form. However, the full conditional PDFs can be written quite easily. Drop et al. [27] used the software package $MART to generate the moments of the posterior PDF, which can be used to infer υ. Alternatively, one can use the MCMC technique to generate the Markov chain and using the chain the BEs of the unknown parameters and the associated CRIs can be computed.

Drop and Mazzuchi [118] considered an ALT where each test item is exposed to several stress levels. The times over which an item is exposed to a specific stress level are different for different items. The order of the stress levels that are applied on an item is also different for different items, even the same stress level also can be repeated. The authors assumed that the failure times are known and the data are Type-I censored. Based on the assumption of the CRM with constant hazard function for each stress level and linear ramping functions, the hazard function under the step-stress pattern is given by Eq. (3.25). One can write the survival function of the lifetime and hence the likelihood function of υ. Drop and Mazzuchi [118] discussed the Bayesian analysis of the model for a multivariate ordered Dirichlet prior on υ based on the MCMC technique.

The classical analysis of the CRM is considered by Kannan et al. [28]. The authors have assumed that there are two stress levels and the complete data are available. It is also assumed that the hazard function of the lifetime are λ_1 and λ_2 at the stress levels s_1 and s_2, respectively. The distribution function of the lifetime under the CRM is given by Eq. (3.25) with $k = 2$. The CDF of the lifetime is given by

$$F(t) = \begin{cases} 0 & \text{if } t \le 0 \\ 1 - e^{-\lambda_1 t} & \text{if } 0 < t \le \tau_1 \\ 1 - e^{-\lambda_1 t - \frac{\lambda_2 - \lambda_1}{2\zeta_1}(t-\tau_1)^2} & \text{if } \tau_1 < t \le \tau_1 + \zeta_1 \\ 1 - e^{-\lambda_1(\tau_1+\zeta_1) - \frac{1}{2}(\lambda_2-\lambda_1)\zeta_1 - \lambda_2(t-\tau_1-\zeta_1)} & \text{if } t > \tau_1 + \zeta_1. \end{cases}$$

Consider the parameterization $(\lambda_1, \lambda_2) \rightarrow (a, b)$ defined by $\lambda_1 = a + b\tau_1$ and $\lambda_2 = a + b(\tau_1 + \zeta_1)$. Clearly it is a one-to-one parameterization and hence inference based on (λ_1, λ_2) is equivalent to the inference based on (a, b). Under the new parameterization the CDF of the lifetime is given by

$$F(t) = \begin{cases} 0 & \text{if } t \le 0 \\ 1 - e^{-(a+b\tau_1)t} & \text{if } 0 < t \le \tau_1 \\ 1 - e^{-at - \frac{b}{2}(t^2 + \tau_1^2)} & \text{if } \tau_1 < t \le \tau_1 + \zeta_1 \\ 1 - e^{-(a+b(\tau_1+\zeta_1))t + \frac{1}{2}b\zeta_1(\tau_1+\zeta_1)} & \text{if } t > \tau_1 + \zeta_1. \end{cases}$$

Let $0 < t_{1:n} < t_{2:n} < \cdots < t_{n_1:n} < \tau_1 < t_{n_1+1:n} < \cdots < t_{n_1+n_2:n} < \tau_1 + \zeta_1 < t_{n_1+n_2+1:n} < \cdots < t_{n:n}$ be the observed data, where n_1, n_2, and $n_3 = n - n_1 - n_2$ are the number of failures in the intervals $(0, \tau_1]$, $(\tau_1, \tau_1 + \zeta_1]$, and $(\tau_1 + \zeta_1, \infty)$, respectively. The log-likelihood function of (a, b) can be obtained as

$$l(a, b) = n_1 \ln (a + b\tau_1) - (a + b\tau_1) \sum_{i=1}^{n_1} t_{i:n} + \sum_{n_1+1}^{n_1+n_2} \ln (a + bt_{i:n}) - a \sum_{i=n_1+1}^{n_1+n_2} t_{i:n}$$

$$- \frac{b}{2} \sum_{i=n_1}^{n_1+n_2} \left(t_{i:n}^2 + \tau_1^2 \right) + n_3 \ln (a + b(\tau_1 + \zeta_1))$$

$$- (a + b(\tau_1 + \zeta_1)) \sum_{i=n_1+n_2+1}^{n} t_{i:n} + \frac{1}{2} n_3 b \zeta_1 (\tau_1 + \zeta_1).$$

It is clear from the log-likelihood function that the MLEs of a and b exist if $n_2 > 0$. If $n_2 = 0$, the MLEs of a and b exist if $n_1 > 0$ and $n_3 > 0$. If $n_2 = 0$ and either of n_1 or n_3 equals zero, the MLEs of the parameters do not exist. When the MLEs exist, $\hat{a}$, the MLE of a, is given by

$$\hat{a} = \frac{n - \hat{b}K}{T},$$

and $\hat{b}$, the MLE of b, is a solution of the following equation.

$$\frac{n_1}{n - b(K - T\tau_1)} + \sum_{i=n_1+1}^{n_1+n_2} \frac{1}{n - b(K - Tt_i)} + \frac{n_3}{n - b(K - T(\tau_1 + \zeta_1))} = 1. \quad (3.28)$$

Here $T = \sum_{i=1}^{n} t_{i:n}$ and

$$K = \tau_1 \sum_{i=1}^{n_1} t_{i:n} + \frac{1}{2} \sum_{i=n_1+1}^{n_1+n_2} \left(t_{i:n}^2 + \tau_1^2 \right) + (\tau_1 + \zeta_1) \sum_{i=n_1+n_2+1}^{n} t_{i:n} + \frac{1}{2} n_3 \zeta_1 (\tau_1 + \zeta_1).$$

Let b_i, $i = 1, 2, \ldots, n_2 + 2$, be such that

$$n - b_1 (K - T\tau_1) = 0,$$

$$n - b_{n_2+2} (K - T(\tau_1 + \zeta_1)) = 0,$$

$$n - b_i (K - Tt_{i+n_1-1:n}) = 0, \quad i = 2, 3, \ldots, n_2 + 1.$$

Then $b_1 > b_2 > \cdots > b_{n_2+1}$. Denoting the righthand side of Eq. (3.28) by $h(b)$, $\lim_{b \to b_i-} h(b) = +\infty$ and $\lim_{b \to b_i+} h(b) = -\infty$. Hence, $h(\cdot)$ has a discontinuity of the second kind at the points $b_1, b_2, \ldots, b_{n_2+2}$. Also $h(b) \downarrow 0$ as $b \to -\infty$, $h(b) \uparrow 0$

as $b \rightarrow \infty$, and $h(0) = 1$. It is quite clear that $h(\cdot)$ is a strictly increasing function on (b_i, b_{i+1}), $i = 0, 1, \ldots, n_2 + 2$ with $b_0 = 0$ and $b_{n_2+3} = \infty$. Hence $h(b) = 0$ has exactly one solution in the interval (b_i, b_{i+1}), $i = 0, 2, \ldots, n_2 + 1$. Let $\hat{b}_i$, $i = 1, 2, \ldots, n_2 + 2$, be the solutions of the Eq. (3.28) and $\hat{a}_i = \frac{n - \hat{b}_i K}{T}$. Compute the quantities $l_i = l(a_i, b_i)$, $i = 1, 2, \ldots, n_2 + 2$. The MLE of (a, b) is the pair $(\hat{a}_i, \hat{b}_i)$, which corresponds to $l_{max} = \max\{l_1, l_2, \ldots, l_{n_2+1}\}$. The authors performed extensive simulation experiments to observe the behavior of the MLEs, and the performances are quite satisfactory.

Step-stress life tests with multiple failure modes

4

4.1 Introduction

We have discussed different issues related to the step-stress models when there is only one failure mode. However, often in reliability or in survival analysis an experimental unit is exposed to more than one fatal risk factor and it might fail due to one of those causes. For example, one individual may die due to cancer, heart attack or kidney failure, etc. Similarly, in a reliability experiment assemblies of ball bearings can fail due to poor lubrication or material defects, a circuit board fails if any of its joints fail, a semiconductor device fails due to either lead or junction failures, etc.

In practice it is usually necessary to assess each risk factor in presence of the other risk factors. In the statistical literature, the problem is known as the competing risks problem, and the analysis is called the competing risks analysis. In a competing risks problem it is as if different risk factors are competing with each other to make the unit fail. For analyzing a competing risks model, each complete observation must be in a bivariate format composed of the failure time and the corresponding cause of failure. To analyze the competing risks data both parametric and nonparametric methods have been used. In the parametric set-up specific lifetime distributions like exponential, Weibull, gamma, log-normal, etc. have been assumed for the different causes, whereas in the nonparametric set-up no specific assumptions on the lifetime distributions are required. To analyze the competing risks data, the lifetime distributions of the different causes can be assumed to be independent or dependent. In most of the situations the lifetime distributions of the different causes are assumed to be independent. Although dependent risk structure might be more realistic, there is an issue about the identifiability of the underlying model; see for example Tsiatis [121]. It has been shown that without the information about the covariates, it is not possible to test the assumption from the data about the independence of the risk factors.

There are two main approaches to analyze competing risks data. One is known as the latent failure time model approach by Cox [122] and the other one is called the cause specific hazard function approach by Prentice et al. [123]. Both the approaches are quite different in nature. It has been shown by Kundu [124] that for the exponential and Weibull lifetimes both the approaches lead to the same likelihood function, hence provide the same set of estimates of the model parameters, although their interpretations are different. An extensive amount of work has been done on the analysis of competing risks data during the last 50 years. See for example two excellent monographs by Crowder [125, 126] which have dealt with the different specific issues related to various competing risks models.

Analysis of Step-Stress Models. http://dx.doi.org/10.1016/B978-0-12-809713-7.00004-1

Klein and Basu [38, 39, 127] first considered the analysis of general accelerated life testing (ALT) models based on competing risks data. The authors used a general stress translation function. Based on the assumptions that the lifetime distributions of the different causes follow exponential or Weibull distributions, inference of the model parameters has been established for a p-component series system, under different censoring schemes. Bai and Chun [128] considered the simple step-stress model and it is assumed that the competing causes of failures have independent exponential distributions. Based on the cumulative exposure model (CEM) assumptions Bai and Chun [128] obtained the optimum simple step-stress model. The results have been extended for the multiple step-stress model and when the competing causes of failures have a mixture of two distributions, by Kim and Bai [129]. Pascual [40, 41] discussed different optimality issues of the ALT models when the competing causes of failures follow Weibull distributions.

Balakrishnan and Han [42] obtained the exact inference of a simple step-stress model when the competing causes of failures follow independent exponential distributions. Based on the CEM assumptions the maximum likelihood estimates (MLEs) of the unknown parameters and their exact distributions have been established when the data are Type-II censored. The method has been extended for Type-I censored data by Han and Balakrishnan [43]. Han and Kundu [45] extended the results of Han and Balakrishnan [43] for the generalized exponential distribution. Srivastava et al. [130] considered the inference of the competing risks data for the Khamis-Higgins model as discussed in Chapter 1. The results have been extended for the progressive censoring scheme and for the hybrid Type-I progressive censoring scheme by Lu and Shi [131] and Zhang et al. [132], respectively. Ganguly and Kundu [133] considered the same problem as in Balakrishnan and Han [42], but it is assumed that the stress changing time is random.

Basu and Ghosh [134] introduced the complementary risks model, which is very close to the competing risks model. In a complementary risks problem, one observes the maximum of the latent failure lifetime instead of the minimum, as it is observed in a competing risks problem. It can be thought of as the lifetime of the last remaining duplicated organ, say for example, kidney. It is the time to failure of a parallel system, but in this case the component due to which the system has finally failed is also observed. Several issues are common between the competing and complementary risks models. Han [46] considered the complementary risks model when the lifetime distributions of the different causes follow the generalized exponential distribution and provided an interesting application of this model.

The rest of the chapter has been arranged as follows. In Section 4.2 we discuss the latent failure time model and show how it has been used in step-stress modeling. The exact inference related to the exponential distribution based on the CEM assumptions is provided in Section 4.3. In Section 4.4 we discuss the inference procedures for the parameters of the exponential distribution based on the cumulative risk model (CRM) assumptions and in Section 4.5 we consider when the competing causes of failures follow Weibull distributions and satisfy the tampered failure rate model (TFRM) assumptions. Finally, the complementary risks model is discussed in Section 4.6.

4.2   SSLT in the presence of competing risks

In reliability analysis Cox's [122] latent failure time model approach has been used quite extensively to analyze competing risks data. In this section we describe briefly the latent failure time model and show how it can be used in step-stress modeling. For notational simplicity, in all the analyses, it is assumed that there are only two causes of failures and it is a simple step-stress model, although all the results can be generalized for more than two causes of failures and for the multiple step-stress model.

4.2.1   Latent failure time model

In a competing risks set-up the observed outcome comprises T, the time to failure, and Δ, the cause of failure. Here we assume that T is an absolutely continuous random variable and Δ is a discrete random variable taking values $1, \ldots, p$, the number of possible causes of failures. In a latent failure time model, the following assumptions have been made. Suppose $X_1, \ldots, X_p$ are p nonnegative random variables corresponding to p causes; then

$$T = \min\{X_1, \ldots, X_p\}.$$

It is assumed that $X_1, \ldots, X_p$ are independently distributed. Therefore, if $\min\{X_1, \ldots, X_p\} = X_j$, for $1 \le j \le p$, then $(T = X_j, \Delta = j)$.

Suppose the random variables $X_1, \ldots, X_p$ have absolute continuous distribution with cumulative distribution functions (CDFs) $F_1(t), \ldots, F_p(t)$, and the associated probability density functions (PDFs) are $f_1(t), \ldots, f_p(t)$, respectively. It is further assumed that $h_1(t), \ldots, h_p(t)$ are the hazard functions of $X_1, \ldots, X_p$, respectively.

Now based on the assumption that $X_1, \ldots, X_p$ are independently distributed, the joint PDF of T and Δ for $t > 0$ and $j = 1, \ldots, p$ can be written as follows:

$$f_{T,\Delta}(t,j) = f_j(t) \prod_{\substack{i=1 \\ i \ne j}}^{p} (1 - F_i(t)) = \frac{f_j(t)}{1 - F_j(t)} \prod_{i=1}^{p} (1 - F_i(t)) = h_j(t) P(T > t).$$

Hence, the marginal PDF of T for $t > 0$ and the probability mass function (PMF) Δ for $j = 1, \ldots, p$ can be obtained as

$$f_T(t) = \sum_{j=1}^{p} h_j(t) P(T > t) = P(T > t) \sum_{j=1}^{p} h_j(t)$$

and

$$P(\Delta = j) = \int_0^{\infty} h_j(t) P(T > t) dt,$$

respectively. Therefore, the conditional PDF of T given $\Delta = j$ is

$$f_{T|\Delta=j}(t|j) = \frac{f_{T,\Delta}(t,j)}{P(\Delta = j)} = \frac{h_j(t)P(T > t)}{\int_0^\infty h_j(u)P(T > u)du} \tag{4.1}$$

and

$$P(\Delta = j|T = t) = \frac{f_{T,\Delta}(t,j)}{f_T(t)} = \frac{h_j(t)}{\sum_{i=1}^p h_i(t)}. \tag{4.2}$$

For some specific parametric distributions, like exponential or Weibull, Eqs. (4.1), (4.2) can be computed explicitly.

Since the survival function of T for $t > 0$ is

$$P(T > t) = \prod_{i=1}^p (1 - F_i(t)),$$

the hazard function of T becomes

$$h_T(t) = \frac{f_T(t)}{P(T > t)} = h_1(t) + \cdots + h_p(t), \tag{4.3}$$

for $t > 0$. Therefore, from Eq. (4.3) it is clear that the hazard function of the item is a sum of the hazard functions of the individual causes. Now let us consider three different examples.

Example 4.1. Suppose $X_1, \ldots, X_p$ are n independent exponential random variables with parameters $\lambda_1, \ldots, \lambda_p$, respectively, i.e., the PDF of $X_i; i = 1, \ldots, p$ is

$$f_{X_i}(t; \lambda_i) = \begin{cases} 0 & \text{if } t < 0 \\ \lambda_i e^{-\lambda_i t} & \text{if } 0 \le t < \infty. \end{cases}$$

Hence, the joint PDF of T and Δ for $t > 0$ and for $j = 1, \ldots, p$ becomes

$$f_{T,\Delta}(t,j) = \lambda_j e^{-(\sum_{i=1}^p \lambda_i)t}.$$

In this case the marginals of T and Δ become

$$f_T(t) = \left(\sum_{i=1}^p \lambda_i\right) e^{-(\sum_{i=1}^p \lambda_i)t}; \quad t > 0$$

and

$$P(\Delta = j) = \int_0^\infty \lambda_j e^{-(\sum_{i=1}^p \lambda_i)t} dt = \frac{\lambda_j}{\sum_{i=1}^p \lambda_i}; \quad j = 1, \ldots, p,$$

respectively.

Observe that the marginal distribution of T is also an exponential distribution with the parameter $\lambda_1 + \cdots + \lambda_p$. The conditional PDF of T given $\Delta = j$ is

$$f_{T|\Delta=j}(t|j) = \frac{f_{T,\Delta}(t,j)}{P(\Delta = j)} = \left(\sum_{i=1}^{p} \lambda_i\right) e^{-(\sum_{i=1}^{p} \lambda_i)t}; \quad t > 0.$$

Example 4.2. Suppose $X_1, \ldots, X_p$ are independent Weibull random variables with the same shape parameter α and different scale parameters $\lambda_1, \ldots, \lambda_p$, respectively, and the PDF of X_i, for $i = 1, \ldots, p$ is

$$f_{X_i}(t; \alpha, \lambda_i) = \begin{cases} 0 & \text{if} \quad t < 0 \\ \alpha \lambda_i t^{\alpha-1} e^{-\lambda_i t^\alpha} & \text{if} \quad 0 \le t < \infty. \end{cases}$$

The joint PDF of T and Δ for $t > 0$ and $j = 1, \ldots, p$ becomes

$$f_{T,\Delta}(t,j) = \alpha \lambda_j t^{\alpha-1} e^{-(\sum_{i=1}^{p} \lambda_i)t^\alpha}.$$

The marginals of T and Δ are

$$f_T(t) = \alpha \left(\sum_{i=1}^{p} \lambda_i\right) t^{\alpha-1} e^{-(\sum_{i=1}^{p} \lambda_i)t^\alpha}; \quad t > 0$$

and

$$P(\Delta = j) = \int_0^\infty \alpha \lambda_j t^{\alpha-1} e^{-(\sum_{i=1}^{p} \lambda_i)t^\alpha} \, dt = \frac{\lambda_j}{\sum_{i=1}^{p} \lambda_i}; \quad j = 1, \ldots, p,$$

respectively. Therefore, the conditional PDF of T given $\Delta = j$ is

$$f_{T|\Delta=j}(t) = \frac{f_{T,\Delta}(t,j)}{P(\Delta = j)} = \alpha \left(\sum_{i=1}^{p} \lambda_i\right) t^{\alpha-1} e^{-(\sum_{i=1}^{p} \lambda_i)t^\alpha}; \quad t > 0.$$

Example 4.3. Suppose $X_1, \ldots, X_p$ are p independent generalized exponential random variables with shape parameters $\alpha_1, \ldots, \alpha_p$ and scale parameters $\lambda_1, \ldots, \lambda_p$, respectively, and the PDF of X_i is

$$f_{X_i}(t; \alpha_i, \lambda_i) = \begin{cases} 0 & \text{if} \quad t < 0 \\ \alpha_i \lambda_i e^{-\lambda_i t}(1 - e^{-\lambda_i t})^{\alpha_i - 1} & \text{if} \quad 0 \le t < \infty. \end{cases}$$

The hazard function of a generalized exponential distribution is

$$h_j(t) = \frac{\alpha_j \lambda_j e^{-\lambda_j t}(1 - e^{-\lambda_j t})^{\alpha_j - 1}}{1 - (1 - e^{-\lambda_j t})^{\alpha_j}}; \quad t > 0. \tag{4.4}$$

The joint PDF of T and Δ, for $t > 0$, and $j = 1, \ldots, p$ is

$$f_{T,\Delta}(t,j) = \frac{\alpha_j \lambda_j e^{-\lambda_j t} (1 - e^{-\lambda_j t})^{\alpha_j - 1}}{1 - (1 - e^{-\lambda_j t})^{\alpha_j}} \times \prod_{i=1}^{p} \left[1 - (1 - e^{-\lambda_i t})^{\alpha_i} \right].$$

Therefore, the marginals of T and Δ are

$$f_T(t) = \prod_{i=1}^{p} \left[1 - (1 - e^{-\lambda_i t})^{\alpha_i} \right] \times \sum_{j=1}^{p} \frac{\alpha_j \lambda_j e^{-\lambda_j t} (1 - e^{-\lambda_j t})^{\alpha_j - 1}}{1 - (1 - e^{-\lambda_j t})^{\alpha_j}},$$

and

$$P(\Delta = j) = \frac{h_j(t)}{\sum_{i=1}^{p} h_i(t)},$$

respectively. Here $h_j(t)$ for $j = 1, \ldots, p$ is the same as defined in Eq. (4.4).

4.2.2 CEM for competing risks

In this section we illustrate how the CEM for a simple step-stress life test (SSLT) can be extended in the presence of competing risks. It is assumed that the stress is changed at a fixed time τ_1 from s_1 to s_2 and we have only two causes of failures, i.e., $p = 2$. We make the following assumptions in this case. It is assumed that when the stress is changed at the time τ_1, the distribution functions of both the latent failures change and both of them follow CEM assumptions. Therefore, the CDF of the latent failure lifetime X_j due to cause j for $j = 1$ and 2 is given by

$$F_j(t) = \begin{cases} F_{j1}(t) & \text{if } t < \tau_1 \\ F_{j2}(t - a_j) & \text{if } t \geq \tau_1. \end{cases}$$

Here F_{j1} and F_{j2} are the CDFs of the latent failure lifetime X_j under the stress levels s_1 and s_2, respectively, and a_j is such that it satisfies

$$F_{j2}(\tau_1 - a_j) = F_{j1}(\tau_1). \tag{4.5}$$

Therefore, if $f_{j1}(t)$ and $f_{j2}(t)$ denote the PDFs of $F_{j1}(t)$ and $F_{j2}(t)$, respectively, the PDF of X_j can be written as

$$f_j(t) = \begin{cases} f_{j1}(t) & \text{if } t < \tau_1 \\ f_{j2}(t - a_j) & \text{if } t \geq \tau_1. \end{cases}$$

Since $T = \min\{X_1, X_2\}$, the CDF of T is

$$F_T(t) = 1 - (1 - F_1(t))(1 - F_2(t))$$

$$= \begin{cases} 1 - (1 - F_{11}(t))(1 - F_{21}(t)) & \text{if } t < \tau_1 \\ 1 - (1 - F_{12}(t - a_1))(1 - F_{22}(t - a_2)) & \text{if } t \geq \tau_1 \end{cases} \quad (4.6)$$

and the corresponding PDF becomes

$$f_T(t) = f_1(t)(1 - F_2(t)) + f_2(t)(1 - F_1(t))$$

$$= \begin{cases} f_{11}(t)(1 - F_{21}(t)) + f_{21}(t)(1 - F_{11}(t)) & \text{if } t < \tau_1 \\ f_{12}(t - a_1)(1 - F_{22}(t - a_2)) + f_{22}(t - a_2)(1 - F_{12}(t - a_1)) & \text{if } t \geq \tau_1. \end{cases}$$
$$(4.7)$$

Therefore, the joint PDF of T and Δ becomes

$$f_{T,\Delta}(t, j) = f_j(t)(1 - F_{j'}(t)) = \begin{cases} f_{j1}(t)(1 - F_{j'1}(t)) & \text{if } t < \tau_1 \\ f_{j2}(t - a_j)(1 - F_{j'2}(t - a_{j'})) & \text{if } t \geq \tau_1 \end{cases}$$

for $j, j' = 1, 2$ and $j \neq j'$.

In a step-stress experiment with competing causes of failures one observes the failure time as well as the cause of failure. The following assumptions have been made on the model. Suppose n items are put on a life testing experiment and the stress is changed at τ_1. Let r be a prefixed positive integer, which is less than n. The experiment continues till r units fail. Each experimental unit can fail for exactly one of the two causes. The data will be of the following form:

$$\mathcal{D} = \{(t_{1:n}, \delta_1), (t_{2:n}, \delta_2), \ldots, (t_{r:n}, \delta_r)\}, \quad (4.8)$$

where

$$t_{1:n} < t_{2:n} < \cdots < t_{n_1:n} < \tau_1 < t_{n_1+1:n} < \cdots < t_{r:n},$$

and n_1 denotes the total number of units that fail before τ_1. We further use the following notations for $j = 1$ and 2:

n_{1j} = the number of units that fails before τ_1 due to the risk factor j

n_{2j} = the number of units that fails after τ_1 due to the risk factor j.

Hence, $n_1 = n_{11} + n_{12}$ and $r - n_1 = n_{21} + n_{22}$. If it is assumed that the lifetime distribution of T depends on the parameter θ, then the likelihood function of θ based on the observed data (4.8) can be written as

$$L(\theta) = \frac{n!}{(n-r)!} \left\{ \prod_{i=1}^{n_1} f_{T,\Delta}(t_{i:n}, \delta_i) \right\} \left\{ \prod_{i=n_1+1}^{r} f_{T,\Delta}(t_{i:n}, \delta_i) \right\} (1 - F_T(t_{r:n}))^{n-r}.$$
$$(4.9)$$

In the subsequent sections we consider some specific cases and discuss their inferential procedures.

4.2.3 TFRM for competing risks

In the last subsection we have shown how the CEM can be extended for competing risks data. The main aim of this subsection is to illustrate how the TFRM of Bhattacharyya and Soejoeti [21] can be used in the presence of competing risks. We are using the same notations as in Section 4.2.2. It is also assumed here that we have only two failure modes and two stress factors. We are using the notations $h_{j1}(t)$, $h_{j2}(t)$ as the hazard function associated with $f_{j1}(t)$ and $f_{j2}(t)$, respectively, for $j = 1$ and 2. Now, if we denote $h_j(t)$ as the overall hazard function of the latent failure time X_j, due to cause j, then due to the TFRM assumption for each cause we have

$$
h_j(t) = \begin{cases} h_{j1}(t) & \text{if } t < \tau_1 \\ h_{j2}(t) = \beta_j h_{j1}(t) & \text{if } t \ge \tau_1, \end{cases}
$$

for some $\beta_j > 0$, for $j = 1$ and 2. Hence, the distribution function F_j due to cause j can be written as

$$
\begin{aligned}
F_j(t) &= 1 - \exp\left(-\int_0^t h_j(u)du\right) \\
&= \begin{cases} F_{j1}(t) = 1 - \exp\left(-\int_0^t h_{j1}(u)du\right) & \text{if } t < \tau_1 \\ F_{j2}(t) = 1 - \exp\left(-\int_0^{\tau_1} h_{j1}(u)du - \beta_j \int_{\tau_1}^t h_{j1}(u)du\right) & \text{if } t \ge \tau_1, \end{cases}
\end{aligned}
$$

and the associated PDF becomes

$$
\begin{aligned}
f_j(t) &= h_j(t)\exp\left(-\int_0^t h_j(u)du\right) \\
&= \begin{cases} f_{j1}(t) = h_{j1}(t)\exp\left(-\int_0^t h_{j1}(u)du\right) & \text{if } t < \tau_1 \\ f_{j2}(t) = \beta_j h_{j1}(t)\exp\left(-\int_0^{\tau_1} h_{j1}(u)du - \beta_j \int_{\tau_1}^t h_{j1}(u)du\right) & \text{if } t \ge \tau_1. \end{cases}
\end{aligned}
$$

The CDF of $T = \min\{X_1, X_2\}$ and the associated PDF can be obtained similarly as in Eqs. (4.6) and (4.7), respectively. Hence, the joint PDF of T and Δ becomes

$$
f_{T,\Delta}(t,j) = f_j(t)(1 - F_{j'}(t)) = \begin{cases} f_{j1}(t)(1 - F_{j'1}(t)) & \text{if } t < \tau_1 \\ f_{j2}(t)(1 - F_{j'2}(t)) & \text{if } t \ge \tau_1 \end{cases}
$$

for $j, j' = 1, 2$ and $j \neq j'$. Therefore, based on the data (4.8) and using the same notations as in Section 4.2.2, the likelihood function of θ will be of the same form as Eq. (4.9).

4.2.4 CRM for competing risks

In this section we illustrate how the CRM of Kannan et al. [28] can be extended in the presence of competing risks. Here we will be using the same notations as in Sections 4.2.2 and 4.2.3. Under the CRM, it is assumed that the overall hazard function due to cause j is of the following form:

$$h_j(t) = \begin{cases} h_{j1}(t) & \text{if } t < \tau_1 \\ a_j + b_j t & \text{if } \tau_1 \leq t < \tau_1 + \delta \\ h_{j2}(t) & \text{if } t \geq \tau_1 + \delta. \end{cases}$$

Here it is assumed that $h_{j1}(t)$ and $h_{j2}(t)$ are continuous functions, and $a_j > 0$, $b_j > 0$ satisfy the following conditions;

$$a_j + b_j \tau_1 = h_{j1}(\tau_1) \quad \text{and} \quad a_j + b_j(\tau_1 + \delta) = h_{j2}(\tau_1 + \delta); \quad j = 1, 2.$$

Hence, the CDF F_j due to cause j can be written as

$$F_j(t) = 1 - \exp\left(-\int_0^t h_j(u)du\right) = \begin{cases} F_{j1}(t) & \text{if } t < \tau_1 \\ F_{jm}(t) & \text{if } \tau_1 \leq t < \tau_1 + \delta \\ F_{j2}(t) & \text{if } t \geq \tau_1 + \delta, \end{cases}$$

where

$$F_{j1}(t) = 1 - \exp\left(-\int_0^t h_{j1}(u)du\right),$$

$$F_{jm}(t) = 1 - \exp\left(-\int_0^{\tau_1} h_{j1}(u)du - a_j(t - \tau_1) - \frac{b_j}{2}(t^2 - \tau_1^2)\right),$$

$$F_{j2}(t) = 1 - \exp\left(-\int_0^{\tau_1} h_{j1}(u)du - a_j\delta - \frac{b_j}{2}((\tau_1 + \delta)^2 - \tau_1^2) - \int_{\tau_1}^t h_{j2}(u)du\right),$$

and the associated PDF becomes

$$f_j(t) = h_j(t) \exp\left(-\int_0^t h_j(u)du\right) = \begin{cases} f_{j1}(t) & \text{if } t < \tau_1 \\ f_{jm}(t) & \text{if } \tau_1 \leq t < \tau_1 + \delta \\ f_{j2}(t) & \text{if } t \geq \tau_1 + \delta, \end{cases}$$

where

$$f_{j1}(t) = h_{j1}(t) \exp\left(-\int_0^t h_{j1}(u)du\right),$$

$$f_{jm}(t) = (a_j + b_j t) \exp\left(-\int_0^{\tau_1} h_{j1}(u)du - a_j(t-\tau_1) - \frac{b_j}{2}(t^2 - \tau_1^2)\right),$$

$$f_{j2}(t) = h_{j2}(t) \exp\left(-\int_0^{\tau_1} h_{j1}(u)du - a_j\delta - \frac{b_j}{2}((\tau_1+\delta)^2 - \tau_1^2) - \int_{\tau_1}^t h_{j2}(u)du\right).$$

4.3 Exponential distribution: CEM

Balakrishnan and Han [42] first considered the exact inference of the model parameters of the competing risks model in a simple step-stress set-up when the competing causes of failures have exponential distributions, and the data are Type-II censored. They obtained the exact distribution of the MLEs and based on the exact distribution, the confidence intervals of the unknown parameters can be obtained. In a subsequent paper the authors Han and Balakrishnan [43] further considered the exact inference of the unknown parameters when the data are Type-I censored. In this section we provide the necessary results for Type-II censored data.

The following assumptions have been made on the model parameters. Each unit fails by one of the two competing causes, and the time-to-failure of each competing risk has an independent exponential distribution that satisfies the CEM assumption. It is assumed that θ_{ij} is the mean time-to-failure of a test unit at the stress level s_i by the risk factor j, for $i, j = 1, 2$. Therefore, the CDF of the lifetime of the jth cause for $j = 1, 2$ becomes

$$F_j(t) = F_j(t; \theta_{1j}, \theta_{2j}) = \begin{cases} 0 & \text{if } t < 0 \\ 1 - \exp\left\{-\frac{1}{\theta_{1j}}t\right\} & \text{if } 0 \le t < \tau_1 \\ 1 - \exp\left\{-\frac{1}{\theta_{1j}}\tau_1 - \frac{1}{\theta_{2j}}(t-\tau_1)\right\} & \text{if } \tau_1 \le t < \infty, \end{cases}$$

and the corresponding PDF is given by

$$f_j(t) = f_j(t; \theta_{1j}, \theta_{2j}) = \begin{cases} 0 & \text{if } t < 0 \\ \frac{1}{\theta_{1j}} \exp\left\{-\frac{1}{\theta_{1j}}t\right\} & \text{if } 0 \le t < \tau_1 \\ \frac{1}{\theta_{2j}} \exp\left\{-\frac{1}{\theta_{1j}}\tau_1 - \frac{1}{\theta_{2j}}(t-\tau_1)\right\} & \text{if } \tau_1 \le t < \infty. \end{cases}$$

Therefore, if we denote $\boldsymbol{\theta} = (\theta_{11}, \theta_{12}, \theta_{21}, \theta_{22})$, then the CDF of $T = \min\{X_1, X_2\}$ can be obtained as

$$F(t) = F(t; \boldsymbol{\theta}) =$$
$$\begin{cases} 0 & \text{if } t < 0 \\ 1 - \exp\left\{-\left(\frac{1}{\theta_{11}} + \frac{1}{\theta_{12}}\right)t\right\} & \text{if } 0 \le t < \tau_1 \\ 1 - \exp\left\{-\left(\frac{1}{\theta_{11}} + \frac{1}{\theta_{12}}\right)\tau_1 - \left(\frac{1}{\theta_{21}} + \frac{1}{\theta_{22}}\right)(t-\tau_1)\right\} & \text{if } \tau_1 \le t < \infty, \end{cases}$$

and the corresponding PDF is

$$
f(t) = f(t; \boldsymbol{\theta}) =
\begin{cases}
0 & \text{if } t < 0 \\
\left(\frac{1}{\theta_{11}} + \frac{1}{\theta_{12}} \right) \exp\left\{ -\left(\frac{1}{\theta_{11}} + \frac{1}{\theta_{12}} \right) t \right\} & \text{if } 0 \le t < \tau_1 \\
\left(\frac{1}{\theta_{21}} + \frac{1}{\theta_{22}} \right) \exp\left\{ -\left(\frac{1}{\theta_{11}} + \frac{1}{\theta_{12}} \right) \tau_1 - \left(\frac{1}{\theta_{21}} + \frac{1}{\theta_{22}} \right) (t - \tau_1) \right\} & \text{if } \tau_1 \le t < \infty.
\end{cases}
$$

Therefore, in this case the joint PDF of T and Δ takes the following form:

$$
f_{T,\Delta}(t,j) = f_{T,\Delta}(t; \boldsymbol{\theta}) =
\begin{cases}
0 & \text{if } t < 0 \\
\frac{1}{\theta_{1j}} \exp\left\{ -\left(\frac{1}{\theta_{11}} + \frac{1}{\theta_{12}} \right) t \right\} & \text{if } 0 \le t < \tau_1 \\
\frac{1}{\theta_{2j}} \exp\left\{ -\left(\frac{1}{\theta_{11}} + \frac{1}{\theta_{12}} \right) \tau_1 - \left(\frac{1}{\theta_{21}} + \frac{1}{\theta_{22}} \right) (t - \tau_1) \right\} & \text{if } \tau_1 \le t < \infty.
\end{cases}
\tag{4.10}
$$

4.3.1 Conditional MLEs and their distributions

Based on the Type-II censored competing risks data as in Eq. (4.8), using the joint PDF of T and Δ as in Eq. (4.10), the likelihood function of $\boldsymbol{\theta}$ can be written as

$$
L(\boldsymbol{\theta}) = \frac{n!}{(n-r)!} \left\{ \prod_{i,j=1}^{2} \theta_{ij}^{-n_{ij}} \right\} \exp\left\{ -\left(\frac{1}{\theta_{11}} + \frac{1}{\theta_{12}} \right) U_1 - \left(\frac{1}{\theta_{21}} + \frac{1}{\theta_{22}} \right) U_2 \right\},
\tag{4.11}
$$

for $t_{1:n} < \cdots < t_{n_1:n} < \tau_1 < t_{n_1+1:n} < \cdots < t_{r:n} < \infty$. Here

$$
U_1 = \sum_{i=1}^{n_1} t_{i:n} + (n - n_1)\tau_1,
$$

$$
U_2 = \sum_{i=n_1+1}^{r} (t_{i:n} - \tau_1) + (n - r)(t_{r:n} - \tau_1),
$$

and n_{ij}'s and n_1 are the same as defined in Section 4.2.2. First observe that U_1 and U_2 denote the total time on test at the stress level s_1 and s_2, respectively. It is clear from the likelihood function (4.11) that the MLEs of θ_{ij} for all $i, j = 1, 2$ exist provided $n_{ij} \ge 1$ for all $i, j = 1, 2$. Hence, if any of the $n_{ij} = 0$ for $i, j = 1, 2$, then the MLE of the corresponding θ_{ij} does not exist. From now onward, we make this assumption that $n_{ij} \ge 1$, for $i, j = 1, 2$. Based on this assumption, the log-likelihood function without the additive constant can be written as

$$l(\boldsymbol{\theta}) = \sum_{i,j=1}^{2} n_{ij} \ln \theta_{ij} - \left(\frac{1}{\theta_{11}} + \frac{1}{\theta_{12}}\right) U_1 - \left(\frac{1}{\theta_{21}} + \frac{1}{\theta_{22}}\right) U_2.$$

Hence, the MLEs of θ_{ij} can be easily obtained as

$$\hat{\theta}_{11} = \frac{U_1}{n_{11}}, \quad \hat{\theta}_{12} = \frac{U_1}{n_{12}}, \quad \hat{\theta}_{21} = \frac{U_2}{n_{21}}, \quad \hat{\theta}_{22} = \frac{U_2}{n_{22}}. \tag{4.12}$$

Comment 4.1

In a competing risks set-up in many situations it is known that a relationship exists between the mean failure times due to different causes of failures. Therefore, it may be reasonable to assume that there exist known $0 < \rho_1, \rho_2 < 1$ such that $\theta_{21} = \rho_1 \theta_{11}$, and $\theta_{22} = \rho_2 \theta_{12}$. In this case the MLEs of all the unknown parameters exist provided $n_{11} + n_{21} > 0$ and $n_{12} + n_{22} > 0$, and the MLEs are as follows:

$$\hat{\theta}_{11} = \frac{1}{n_{11} + n_{21}} \left[U_1 + \frac{1}{\rho_1} U_2\right], \quad \hat{\theta}_{21} = \rho_1 \hat{\theta}_{11},$$

$$\hat{\theta}_{12} = \frac{1}{n_{12} + n_{22}} \left[U_1 + \frac{1}{\rho_2} U_2\right], \quad \hat{\theta}_{22} = \rho_2 \hat{\theta}_{12}.$$

Comment 4.2

The simple step-stress model can be easily generalized to the multiple step-stress model along the same manner. Moreover, in this case if $\theta_{11}, \theta_{21} \to \infty$ or $\theta_{12}, \theta_{22} \to \infty$, then the preceding model reduces to the simple step-stress model without the competing risks. If $\tau_1 \to \infty$, it reduces to the standard Type-II censored competing risks model.

Balakrishnan and Han [42] obtained the exact conditional distributions of $\hat{\theta}_{ij}$ as provided in Eq. (4.12) conditioning on $N_{ij} \geq 1$, for all $i, j = 1, 2$. They first obtained the conditional moment generating function (CMGF) of $\hat{\theta}_{ij}$, and then by inverting it obtained the exact conditional distribution. We provide the main results; the detailed proofs can be found in Balakrishnan and Han [42]. We need to use the following notations to state the main results.

Let us denote $\boldsymbol{N} = (N_{11}, N_{12}, N_{21}, N_{22})$, and let $\boldsymbol{n}$ be the observed integer vector of $\boldsymbol{N}$. Let

$$\mathcal{N} = \left\{ \boldsymbol{n} \,\middle|\, n_{ij} \geq 1, i, j = 1, 2, \text{ and } \sum_{i,j=1}^{2} n_{ij} = r \right\}.$$

Further, the joint probability mass function of N is given by

$$P(N = n) = \binom{n}{n_{11}, n_{12}} \binom{n_{21} + n_{22}}{n_{21}} \left\{ \prod_{i,j=1}^{2} \pi_{ij}^{n_{ij}} \right\} \{F(\tau_1)\}^{n_{11}+n_{12}} \{1 - F(\tau_1)\}^{n-n_{11}-n_{12}},$$

where for integers n, m_1 and m_2 satisfying $m_1 + m_2 < n$, $\binom{n}{m_1, m_2} = \frac{n!}{m_1! m_2! (n - m_1 - m_2)!}$. Here, π_{1j} and π_{2j} denote the relative risk imposed on a test unit before and after τ_1, respectively, due to the risk factor j, for $j = 1, 2$, and they are

$$\pi_{1j} = P(\Delta = j | 0 < T < \tau_1) = \frac{\theta_{1j}^{-1}}{\theta_{11}^{-1} + \theta_{12}^{-1}} \quad \text{and}$$

$$\pi_{2j} = P(\Delta = j | T \geq \tau_1) = \frac{\theta_{2j}^{-1}}{\theta_{21}^{-1} + \theta_{22}^{-1}}.$$

Further,

$$F(\tau_1) = 1 - \exp\left\{ -\left(\frac{1}{\theta_{11}} + \frac{1}{\theta_{12}} \right) \tau_1 \right\}.$$

Now inverting the CMGF of $\hat{\theta}_{ij}$, conditioning on the event that $\{N \in \mathcal{N}\}$, the conditional PDFs of $\hat{\theta}_{ij}$ can be obtained as follows.

Theorem 4.3.1. *The conditional PDF of $\hat{\theta}_{1j}$, given $N \in \mathcal{N}$, is*

$$f_{\hat{\theta}_{1j}}(x) = f_{\hat{\theta}_{1j}}(x | N \in \mathcal{N}) = \sum_{n \in \mathcal{N}} \sum_{k=0}^{n_{11}+n_{12}} C_{n,jk}^{[1]} f_G\left(x - \tau_{jk}; n_{11} + n_{12}, n_{1j}\left(\frac{1}{\theta_{11}} + \frac{1}{\theta_{12}} \right) \right)$$

for $i, j = 1, 2$, where

$$\tau_{jk} = (n - n_{11} - n_{12} + k) \frac{\tau_1}{n_{1j}},$$

$$C_{n,jk}^{[1]} = \frac{(-1)^k}{\sum_{m \in \mathcal{N}} P(N = m)} \binom{n}{n_{11}, n_{12}} \binom{n_{21} + n_{22}}{n_{21}} \binom{n_{11} + n_{12}}{k}$$

$$\times \left\{ \prod_{i', j'=1}^{2} \pi_{i'j'}^{n_{i'j'}} \right\} \times \exp\left\{ -\left(\frac{1}{\theta_{11}} + \frac{1}{\theta_{12}} \right) \tau_{jk} n_{1j} \right\},$$

$$f_G(y; \alpha, \lambda) = \begin{cases} \frac{\lambda^\alpha}{\Gamma(\alpha)} y^{\alpha-1} e^{-\lambda y} & \text{if } y > 0 \\ 0 & \text{if } y \leq 0, \end{cases}$$

for $\alpha > 0$ and $\lambda > 0$.

Theorem 4.3.2. *The conditional PDF of $\hat{\theta}_{2j}$, given $N \in \mathcal{N}$, is*

$$f_{\hat{\theta}_{2j}}(x) = f_{\hat{\theta}_{2j}}(x|N \in \mathcal{N}) = \sum_{n \in \mathcal{N}} C_n^{[2]} f_G\left(x; n_{21} + n_{22}, n_{2j}\left(\frac{1}{\theta_{21}} + \frac{1}{\theta_{22}}\right)\right)$$

for $i, j = 1, 2$, where

$$C_n^{[2]} = P(N = n|N \in \mathcal{N}) = \frac{P(N = n)}{\sum_{m \in \mathcal{N}} P(N = m)}$$

$$= \frac{1}{\sum_{m \in \mathcal{N}} P(N = m)} \binom{n}{n_{11}, n_{12}} \binom{n_{21} + n_{22}}{n_{21}} \left\{ \prod_{i,j=1}^{2} \pi_{ij}^{n_{ij}} \right\}$$

$$\times \left(1 - \exp\left\{-\left(\frac{1}{\theta_{11}} + \frac{1}{\theta_{12}}\right)\tau_1\right\}\right)^{n_{11}+n_{12}}$$

$$\times \exp\left\{-\left(\frac{1}{\theta_{11}} + \frac{1}{\theta_{12}}\right)\tau_1(n - n_{11} - n_{12})\right\}$$

and $f_G(x; \alpha, \lambda)$ is the same as defined in Theorem 4.3.1.

From Theorems 4.3.1 and 4.3.2, it is clear that the PDF of $\hat{\theta}_{1j}$ is a generalized mixture of shifted gamma distributions, and the PDF of $\hat{\theta}_{2j}$ is a mixture of gamma distribution functions. From the PDFs of $\hat{\theta}_{1j}$, $\hat{\theta}_{2j}$ and using the moments of a gamma distribution function, different moments of the MLEs can be obtained as follows:

$$E[\hat{\theta}_{1j}] = \sum_{n \in \mathcal{N}} \sum_{k=0}^{n_{11}+n_{12}} C_{n,jk}^{[1]} \left\{ \frac{n_{11} + n_{12}}{n_{1j}}\left(\frac{1}{\theta_{11}} + \frac{1}{\theta_{12}}\right)^{-1} + \tau_{jk} \right\},$$

$$E[\hat{\theta}_{1j}^2] = \sum_{n \in \mathcal{N}} \sum_{k=0}^{n_{11}+n_{12}} C_{n,jk}^{[1]} \left\{ \frac{(n_{11} + n_{12})(n_{11} + n_{12} + 1)}{n_{1j}^2}\left(\frac{1}{\theta_{11}} + \frac{1}{\theta_{12}}\right)^{-2} + \tau_{jk}^2 \right.$$

$$\left. + \frac{2(n_{11} + n_{12})}{n_{1j}}\left(\frac{1}{\theta_{11}} + \frac{1}{\theta_{12}}\right)^{-1} \tau_{jk} \right\},$$

$$E[\hat{\theta}_{2j}] = \sum_{n \in \mathcal{N}} C_n^{[2]} \left\{ \frac{n_{21} + n_{22}}{n_{2j}}\left(\frac{1}{\theta_{21}} + \frac{1}{\theta_{22}}\right)^{-1} \right\},$$

$$E[\hat{\theta}_{2j}^2] = \sum_{n \in \mathcal{N}} C_n^{[2]} \left\{ \frac{(n_{21} + n_{22})(n_{21} + n_{22} + 1)}{n_{2j}^2}\left(\frac{1}{\theta_{21}} + \frac{1}{\theta_{22}}\right)^{-2} \right\}.$$

4.3.2 Different confidence intervals

Similarly as shown in Section 2.2.2 it is possible to construct confidence intervals of θ_{ij} for $i, j = 1, 2$ based on the exact distributions of their MLEs. Based on the assumptions that the tail probability of $\hat{\theta}_{ij}$, i.e., $P_\theta(\hat{\theta}_{ij} > a)$ is a strictly increasing function of θ_{ij} for each $i, j = 1, 2$ and for any fixed $a > 0$, the invertibility of the pivotal quantities for the parameter θ_{ij} is guaranteed. Hence, the exact confidence intervals of θ_{1j} and θ_{2j} for $j = 1, 2$ can be obtained as follows. If $\hat{\theta}_{1j}^{\text{obs}}$ denotes the observed value of $\hat{\theta}_{1j}$ and $\theta_{1j}^L, \theta_{1j}^U$ denote the upper and lower limits of a symmetric $100(1 - \alpha)\%$ confidence interval of θ_{1j}, then θ_{1j}^L and θ_{1j}^U can be obtained by solving the following two nonlinear equations:

$$\frac{\alpha}{2} = \sum_{n \in \mathcal{N}} \sum_{k=0}^{n_{11}+n_{12}} C_{n,jk}^{[1]} \Gamma \left(n_{1j} \left(\frac{1}{\theta_{1j}^L} + \frac{1}{\theta_{1j'}} \right) (\hat{\theta}_{1j}^{\text{obs}} - \tau_{1jk}); n_{11} + n_{12} \right),$$

$$1 - \frac{\alpha}{2} = \sum_{n \in \mathcal{N}} \sum_{k=0}^{n_{11}+n_{12}} C_{n,jk}^{[1]} \Gamma \left(n_{1j} \left(\frac{1}{\theta_{1j}^U} + \frac{1}{\theta_{1j'}} \right) (\hat{\theta}_{1j}^{\text{obs}} - \tau_{1jk}); n_{11} + n_{12} \right)$$

for $j = 1, 2$ with $j' \neq j$. Here $\Gamma(x; \alpha)$ is the incomplete gamma function as it has been defined in Section 2.2.2.

Similarly, if $\hat{\theta}_{2j}^{\text{obs}}$ denotes an observed value of $\hat{\theta}_{2j}$ and $\theta_{2j}^L, \theta_{2j}^U$ are the upper and lower limits of a symmetric $100(1 - \alpha)\%$ confidence interval of θ_{2j}, then θ_{2j}^L and θ_{2j}^U can be obtained as the solutions of the following two nonlinear equations:

$$\frac{\alpha}{2} = \sum_{n \in \mathcal{N}} C_n^{[2]} \Gamma \left(n_{2j} \left(\frac{1}{\theta_{2j}^L} + \frac{1}{\theta_{2j'}} \right) \hat{\theta}_{2j}^{\text{obs}}; n_{21} + n_{22} \right),$$

$$1 - \frac{\alpha}{2} = \sum_{n \in \mathcal{N}} C_n^{[2]} \Gamma \left(n_{2j} \left(\frac{1}{\theta_{2j}^U} + \frac{1}{\theta_{2j'}} \right) \hat{\theta}_{2j}^{\text{obs}}; n_{21} + n_{22} \right)$$

for $j = 1, 2$ with $j' \neq j$.

Since the exact confidence intervals are not very easy to obtain, the asymptotic confidence intervals based on the asymptotic distribution of the MLEs can also be used, particularly if the sample size is not too small. In this case the observed Fisher information matrix of $\boldsymbol{\theta}$ is in a simple form as given here:

$$\boldsymbol{I}(\boldsymbol{\theta}) = \begin{bmatrix} \frac{n_{11}}{\hat{\theta}_{11}^2} & 0 & 0 & 0 \\ 0 & \frac{n_{12}}{\hat{\theta}_{12}^2} & 0 & 0 \\ 0 & 0 & \frac{n_{21}}{\hat{\theta}_{21}^2} & 0 \\ 0 & 0 & 0 & \frac{n_{11}}{\hat{\theta}_{22}^2} \end{bmatrix}. \tag{4.13}$$

Hence, based on the Fisher information matrix (4.13), a $100(1 - \alpha)\%$ confidence interval of θ_{ij} can be obtained as

$$\left(\hat{\theta}_{ij} - z_{\alpha/2}\sqrt{V_{ij}}, \hat{\theta}_{ij} + z_{\alpha/2}\sqrt{V_{ij}}\right); \quad i,j = 1,2,$$

where $V_{ij} = \hat{\theta}_{ij}^2/n_{ij}$, for $i,j = 1,2$.

Alternatively, similarly to that shown in Section 2.2.3, the parametric bootstrap method can be used to construct confidence intervals of the unknown parameters in this case. Procedures are very similar to the procedures described in Section 2.2.3 and hence, the details are avoided.

4.3.3 Numerical comparison and data analysis

Balakrishnan and Han [42] performed extensive simulation experiments to observe how the different methods perform in practice. We briefly report their findings as follows. It is observed that the performances of the MLEs are quite satisfactory in the sense that as the sample size increases, the average biases and mean squared errors (MSEs) decrease. The performances of the exact confidence intervals which are obtained by solving nonlinear equations as mentioned previously are very good in the sense they maintain the coverage percentages for all the parameters at different nominal levels and for different τ_1 values, whereas both the asymptotic confidence intervals and the bootstrap confidence intervals cannot maintain the coverage percentages of all the θ_{ij} for all τ_1 values. For example, if τ_1 is small the coverage percentages of θ_{11} and θ_{12} are lower than the nominal value and similarly when τ_1 is large, the corresponding coverage percentages of θ_{21} and θ_{22} are small. It is also observed that when τ_1 is small the expected lengths of the exact confidence intervals for θ_{11} and θ_{12} are significantly larger than the other two. The same phenomenon is observed for θ_{21} and θ_{22} when τ_1 is large.

Now we present the analysis of a data set that was originally reported by Balakrishnan and Han [42]. The data set was generated with $n = 25$, $r = 20$, $\tau_1 = 3$, and $\theta_{11} = 8.96$, $\theta_{12} = 12.18$, $\theta_{21} = 4.48$, $\theta_{22} = 4.06$. The data set is presented in Table 4.1. In this case $n_{11} = 7$, $n_{12} = 5$, $n_{21} = 4$, $n_{22} = 4$. The MLEs of θ_{ij} are provided here:

$$\hat{\theta}_{11} = 7.510, \quad \hat{\theta}_{12} = 10.514, \quad \hat{\theta}_{21} = 4.128, \quad \hat{\theta}_{22} = 4.128.$$

Note that 95% confidence intervals based on different methods are reported in Table 4.2. It is observed that the lengths of the exact confidence intervals of θ_{12}, θ_{21} and θ_{22} are significantly larger than the other two intervals.

4.4   Exponential distribution: CRM

Beltrami [135] in her PhD thesis considered the inference of the unknown parameters of the competing risks model based on the CRM assumptions and when the lifetimes of the competing causes of failures follow the exponential distribution. Suppose λ_{11} and

Table 4.1 **Simulated data set of Balakrishnan and Han [42]**

Stress level 1 (before $\tau_1 = 3$)		Stress level 2 (after $\tau_1 = 3$)	
Failure time	Failure cause	Failure time	Failure cause
0.145	1	3.105	1
0.289	1	3.537	2
0.345	2	3.608	2
0.382	1	3.621	1
0.575	2	3.640	2
0.577	1	3.814	1
1.126	1	4.514	2
1.588	1	4.946	1
1.597	2		
1.772	1		
2.428	2		
2.744	2		

Table 4.2 **95% confidence intervals of θ_{ij} obtained by different methods**

Parameter	Exact	Asymptotic	Bootstrap
$\theta_{11} = 8.96$	(3.647, 16.912)	(1.947, 13.073)	(3.906, 15.606)
$\theta_{12} = 12.18$	(4.635, 28.457)	(1.298, 19.729)	(5.216, 24.401)
$\theta_{21} = 4.48$	(1.633, 19.177)	(0.083, 8.174)	(1.412, 14.185)
$\theta_{22} = 4.06$	(1.633, 19.177)	(0.083, 8.174)	(1.637, 18.534)

λ_{12} are the constant hazards at the initial stress level s_1 for risks 1 and 2, respectively. At the stress level s_2, after the resulting lag period δ, let λ_{21} and λ_{22} represent the constant hazards for risks 1 and 2, respectively. It is assumed that the lifetime of both the causes have the same lag period. Hence, based on the CRM assumptions, the hazard function of the jth cause of failure for $j = 1$ and 2 is given by

$$h_j(t) = \begin{cases} \lambda_{1j} & \text{if } 0 \leq t < \tau_1 \\ (a_j + b_j t) & \text{if } \tau_1 \leq t < \tau_* \\ \lambda_{2j} & \text{if } t \geq \tau_*. \end{cases}$$

Here $\tau_* = \tau_1 + \delta$. To ensure the continuity, a_1, b_1, a_2, b_2 are such that they satisfy the following linear equations:

$$\lambda_{11} = a_1 + b_1\tau_1, \quad \lambda_{12} = a_2 + b_2\tau_1, \quad \lambda_{21} = a_1 + b_1\tau_*, \quad \lambda_{22} = a_2 + b_2\tau_*.$$

Therefore, the survival function and CDF for the jth cause for $j = 1$ and 2 are given by

$$
S_j(t) = \begin{cases}
0 & \text{if} \quad t < 0 \\
e^{-(a_j+b_j\tau_1)t} & \text{if} \quad 0 \le t < \tau_1 \\
e^{-a_jt-\frac{b_j}{2}(\tau_1^2+t^2)} & \text{if} \quad \tau_1 \le t < \tau_* \\
e^{-(a_j+b_j\tau_*)t-\frac{b_j}{2}(\tau_1^2-\tau_*^2)} & \text{if} \quad t > \tau_*
\end{cases}
$$

and

$$
F_j(t) = \begin{cases}
0 & \text{if} \quad t < 0 \\
1 - e^{-(a_j+b_j\tau_1)t} & \text{if} \quad 0 \le t < \tau_1 \\
1 - e^{-a_jt-\frac{b_j}{2}(\tau_1^2+t^2)} & \text{if} \quad \tau_1 \le t < \tau_* \\
1 - e^{-(a_j+b_j\tau_*)t-\frac{b_j}{2}(\tau_1^2-\tau_*^2)} & \text{if} \quad t > \tau_*,
\end{cases}
$$

respectively. The associated PDF for the jth cause for $j = 1$ and 2 is given by

$$
f_j(t) = \begin{cases}
(a_j + b_jt)e^{-(a_j+b_j\tau_1)t} & \text{if} \quad 0 \le t < \tau_1 \\
(a_j + b_jt)e^{-a_jt-\frac{b_j}{2}(\tau_1^2+t^2)} & \text{if} \quad \tau_1 \le t < \tau_* \\
(a_j + b_j\tau_*)e^{-(a_j+b_j\tau_*)t-\frac{b_j}{2}(\tau_1^2-\tau_*^2)} & \text{if} \quad t > \tau_*.
\end{cases}
$$

Based on the step-stress competing risks data one needs to estimate the unknown parameters, namely λ_{11}, λ_{12}, λ_{21}, and λ_{22}. In practice one needs to estimate the unknown lag period δ also. First it is assumed that δ is known, and then maximizing the log-likelihood function, the MLEs of the unknown parameters can be obtained. It may be mentioned that the MLEs of the unknown parameters cannot be obtained explicitly; one needs to use some numerical technique to compute the MLEs. Finally, using the profile likelihood method, the MLE of δ can also be obtained.

Although MLEs of the unknown parameters can be obtained, the distributions of the MLEs cannot be computed in explicit forms as in the CEM. Due to this reason, the asymptotic distribution of the MLEs has been used to construct confidence intervals of the unknown parameters. The parametric bootstrap method can also be used for this purpose. It is observed that the performance of the MLEs is quite satisfactory for moderate sample sizes. For small and moderate sample sizes the bootstrap confidence intervals perform better than the asymptotic confidence intervals in terms of coverage percentages and the length of the confidence intervals. For details, interested readers are referred to the original PhD thesis of Beltrami [135] and also Beltrami [119].

4.5 Weibull distribution: TFRM

In this section we consider the inference procedure of the unknown parameters when the latent failure time distributions are independent Weibull distributions. It is assumed that each unit fails by one of the two competing causes, and time-to-failure of each competing risk has an independent Weibull distribution with different shape and scale parameters. Further, it is assumed that each cause satisfies the TFRM assumption due to stress change.

We are using the same notations as in the previous section. In this case it is assumed that the CDF of the lifetime of the jth cause for $j = 1, 2$ has the following form:

$$F_j(t) = F_j(t; \alpha_j, \theta_{1j}, \theta_{2j}) = \begin{cases} 1 - \exp\left\{-\frac{1}{\theta_{1j}} t^{\alpha_j}\right\} & \text{if } 0 < t < \tau_1 \\ 1 - \exp\left\{-\frac{1}{\theta_{1j}} \tau_1^{\alpha_j} - \frac{1}{\theta_{2j}} (t^{\alpha_j} - \tau_1^{\alpha_j})\right\} & \text{if } t \geq \tau_1 \end{cases}$$

and the corresponding PDF is given by

$$f_j(t) = f_j(t; \alpha_j, \theta_{1j}, \theta_{2j}) = \begin{cases} \frac{\alpha_j}{\theta_{1j}} t^{\alpha_j-1} \exp\left\{-\frac{1}{\theta_{1j}} t^{\alpha_j}\right\} & \text{if } 0 < t < \tau_1 \\ \frac{\alpha_j}{\theta_{2j}} t^{\alpha_j-1} \exp\left\{-\frac{1}{\theta_{1j}} \tau_1 - \frac{1}{\theta_{2j}} (t^{\alpha_j} - \tau_1^{\alpha_j})\right\} & \text{if } t \geq \tau_1. \end{cases}$$

Let $\boldsymbol{\theta} = (\alpha_1, \alpha_2, \theta_{11}, \theta_{12}, \theta_{21}, \theta_{22})$. Now we will show how to obtain the MLE of $\boldsymbol{\theta}$ based on a complete sample:

$$\{(t_{1:n}, \delta_1), (t_{2:n}, \delta_2), \ldots, (t_{n:n}, \delta_n)\},$$

where $t_{1:n} < t_{2:n} < \cdots < t_{n_1:n} < \tau_1 < t_{n_1+1:n} < \cdots < t_{n:n}$. We further use the following notations:

$$I_{11} = \{i : t_{i:n} < \tau_1, \delta_i = 1\}, \quad I_{21} = \{i : t_{i:n} > \tau_1, \delta_i = 1\},$$
$$I_{12} = \{i : t_{i:n} < \tau_1, \delta_i = 2\}, \quad I_{22} = \{i : t_{i:n} > \tau_1, \delta_i = 2\},$$

and n_{ij} denotes the number of elements in the set I_{ij} for $i = 1, 2, \ldots, n, j = 1, 2$. The likelihood function can be written as

$$L(\boldsymbol{\theta}) = n! \left(\prod_{i \in I_{11}} \frac{\alpha_1}{\theta_{11}} t_{i:n}^{\alpha_1-1} e^{-\frac{t_{i:n}^{\alpha_1}}{\theta_{11}} - \frac{t_{i:n}^{\alpha_2}}{\theta_{12}}} \right) \times \left(\prod_{i \in I_{12}} \frac{\alpha_2}{\theta_{12}} t_{i:n}^{\alpha_2-1} e^{-\frac{t_{i:n}^{\alpha_2}}{\theta_{12}} - \frac{t_{i:n}^{\alpha_1}}{\theta_{11}}} \right)$$

$$\times \left(\prod_{i \in I_{21}} \frac{\alpha_1}{\theta_{21}} t_{i:n}^{\alpha_1-1} e^{-\frac{\tau_1^{\alpha_1}}{\theta_{11}} - \frac{(t_{i:n}^{\alpha_1} - \tau_1^{\alpha_1})}{\theta_{21}}} e^{-\frac{\tau_1^{\alpha_2}}{\theta_{12}} - \frac{(t_{i:n}^{\alpha_2} - \tau_1^{\alpha_2})}{\theta_{22}}} \right)$$

$$\times \left(\prod_{i \in I_{22}} \frac{\alpha_2}{\theta_{22}} t_{i:n}^{\alpha_2-1} e^{-\frac{\tau_1^{\alpha_2}}{\theta_{12}} - \frac{(t_{i:n}^{\alpha_2} - \tau_1^{\alpha_2})}{\theta_{22}}} e^{-\frac{\tau_1^{\alpha_1}}{\theta_{11}} - \frac{(t_{i:n}^{\alpha_1} - \tau_1^{\alpha_1})}{\theta_{21}}} \right)$$

$$= n! \left\{ \prod_{i,j=1}^{2} \theta_{ij}^{-n_{ij}} \right\} \alpha_1^{n_{11}+n_{21}} \alpha_2^{n_{12}+n_{22}} \left(\prod_{i \in I_{11} \cup I_{21}} t_{i:n}^{\alpha_1-1} \right) \left(\prod_{i \in I_{12} \cup I_{22}} t_{i:n}^{\alpha_2-1} \right)$$

$$\times \exp\left\{ -\frac{U_{11}(\alpha_1)}{\theta_{11}} - \frac{U_{12}(\alpha_2)}{\theta_{12}} - \frac{U_{21}(\alpha_1)}{\theta_{21}} - \frac{U_{22}(\alpha_2)}{\theta_{22}} \right\}, \quad (4.14)$$

where

$$U_{11}(\alpha_1) = \sum_{i \in I_{11} \cup I_{12}} t_{i:n}^{\alpha_1} + (n - n_1)\tau_1^{\alpha_1},$$

$$U_{12}(\alpha_2) = \sum_{i \in I_{11} \cup I_{12}} t_{i:n}^{\alpha_1} + (n - n_1)\tau_1^{\alpha_2}$$

$$U_{21}(\alpha_1) = \sum_{i \in I_{21} \cup I_{22}} \left(t_{i:n}^{\alpha_1} - \tau_1^{\alpha_1} \right),$$

$$U_{22}(\alpha_2) = \sum_{i \in I_{21} \cup I_{22}} \left(t_{i:n}^{\alpha_2} - \tau_1^{\alpha_2} \right).$$

Therefore, to compute the MLEs of the unknown parameters one needs to maximize Eq. (4.14). It is clear that for the existence of the MLEs we need $n_{ij} > 0$, for $i, j = 1, 2$. Therefore, it is assumed that $n_{ij} > 0$, for $i, j = 1, 2$. Now for a given α_1 and α_2, it can be easily seen that the MLEs of θ_{ij} can be obtained as follows:

$$\hat{\theta}_{11}(\alpha_1) = \frac{U_{11}(\alpha_1)}{n_{11}}, \quad \hat{\theta}_{12}(\alpha_2) = \frac{U_{12}(\alpha_1)}{n_{12}},$$

$$\hat{\theta}_{21}(\alpha_1) = \frac{U_{21}(\alpha_1)}{n_{21}}, \quad \hat{\theta}_{22}(\alpha_2) = \frac{U_{22}(\alpha_1)}{n_{22}}.$$

Hence, the MLEs of α_1 and α_2 can be obtained by maximizing the corresponding profile likelihood functions $g_1(\alpha_1)$ and $g_2(\alpha_2)$, respectively, where

$$g_1(\alpha_1) = \frac{\alpha_1^{n_{11}+n_{21}}}{U_{11}(\alpha_1)^{n_{11}} \times U_{21}(\alpha_1)^{n_{21}}} \prod_{i \in I_{11} \cup I_{21}} t_{i:n}^{\alpha_1 - 1} \quad \text{and} \tag{4.15}$$

$$g_2(\alpha_2) = \frac{\alpha_2^{n_{12}+n_{22}}}{U_{12}(\alpha_2)^{n_{12}} \times U_{22}(\alpha_2)^{n_{22}}} \prod_{i \in I_{12} \cup I_{22}} t_{i:n}^{\alpha_2 - 1}. \tag{4.16}$$

Under some mild restrictions on the data it can be shown that Eqs. (4.15), (4.16) are unimodal functions of α_1 and α_2, respectively; hence the MLEs of α_1 and α_2 exist and they are unique. If we denote the MLEs of α_1 and α_2, as $\hat{\alpha}_1$ and $\hat{\alpha}_2$, respectively, then the MLEs of θ_{ij}, for $i, j = 1$ and 2 become

$$\hat{\theta}_{11} = \hat{\theta}_{11}(\hat{\alpha}_1), \quad \hat{\theta}_{12} = \hat{\theta}_{12}(\hat{\alpha}_2), \quad \hat{\theta}_{21} = \hat{\theta}_{21}(\hat{\alpha}_1), \quad \hat{\theta}_{22} = \hat{\theta}_{22}(\hat{\alpha}_2).$$

Since the MLEs cannot be obtained in explicit forms, it is difficult to derive the exact distribution of the MLEs. Asymptotic normality results can be used to construct asymptotic confidence intervals of the unknown parameters. Alternatively parametric bootstrap methods can be used for this purpose. Recently, Wu et al. [136], Zhang et al. [132], and Liu and Shi [131] provided the analysis of a simple step-stress model under different censoring schemes when the lifetime distributions of the experimental units follow a Weibull distribution.

Open Problem: Although quite a bit of work has been done in developing the inference procedure of a simple step-stress model in the case of Weibull distribution based on the classical approach, no work has been done to develop the Bayesian inference of the unknown parameters. It will be interesting to develop the Bayes inference under this set-up and compare their performances with the classical ones.

4.6 SSLT in the presence of complementary risks

Although extensive work has been done related to the competing risks model, the amount of literature related to the complementary risks problem is rather less. Interestingly, most of the work related to the complementary risks model is related to the identifiability issues of the model; see for example Basu and Kelin [137], Basu [138], and Basu and Ghosh [134]. It can be shown similarly to the competing risks model that the independence of the complementary causes cannot be tested from the observed data. Therefore, in this case also it is quite natural to model the lifetime distribution as follows.

Suppose T denotes the time to failure and Δ denotes the cause of failure as before. Moreover, $X_1, \ldots, X_p$ are p nonnegative random variables corresponding to p causes; then

$$T = \max\{X_1, \ldots, X_p\}.$$

Therefore, if $\max\{X_1, \ldots, X_p\} = X_j$, for $1 \le j \le p$, then $(T = X_j, \Delta = j)$. It is assumed similar to the latent failure time model assumptions of Cox [122] for competing risks model that $X_1, \ldots, X_p$ are independently distributed. Therefore, the joint PDF of T and Δ for $t > 0$, and for $j = 1, \ldots, p$, can be written as

$$f_{T,\Delta}(t,j) = f_j(t) \prod_{\substack{i=1 \\ i \ne j}}^{p} F_i(t) = \frac{f_j(t)}{F_j(t)} \prod_{i=1}^{p} F_i(t) = r_j(t)P(T \le t).$$

Here $r_j(t) = \dfrac{f_j(t)}{F_j(t)}$ is known as the reversed hazard function of X_j. Although, the hazard function has been very well studied in the statistical literature the same is not true in case of the reversed hazard function. Interested readers may see Block et al. [139], Kundu and Gupta [140], and the references cited therein.

The marginals of T for $t > 0$ and Δ for $j = 1, \ldots, p$ can be obtained as

$$f_T(t) = \sum_{j=1}^{p} r_j(t)P(T \le t) = P(T \le t) \sum_{j=1}^{p} r_j(t), \tag{4.17}$$

and

$$P(\Delta = j) = \int_0^{\infty} r_j(t)P(T \le t)dt,$$

respectively. Therefore, the conditional PDF of T given $\Delta = j$ is

$$f_{T|\Delta=j}(t) = \frac{f_{T,\Delta}(t,j)}{P(\Delta = j)} = \frac{r_j(t)P(T \le t)}{\displaystyle\int_0^\infty r_j(t)P(T \le t)dt}$$

and

$$P(\Delta = j|T = t) = \frac{f_{T,\Delta}(t,j)}{f_T(t)} = \frac{r_j(t)}{\sum_{j=1}^p r_j(t)}.$$

From Eq. (4.17) one can easily obtain the reversed hazard rate of the distribution of the lifetime as

$$r_T(t) = \sum_{j=1}^p r_j(t).$$

Now we demonstrate how the CEM can be used in the step-stress set-up in the presence of complementary risks.

4.6.1 CEM for complementary risks

We are making similar assumptions as in the competing risks model. It is assumed that the stress is changed at a fixed time τ_1 from s_1 to s_2, and we have only two causes of failures, i.e., $p = 2$. It is further assumed, as before, that when the stress is changed at τ_1, the CDFs of both the latent failure times change and they follow CEM assumptions. Hence, the CDF and PDF of the latent failure time X_j for $j = 1$ and 2 are the same as in Section 4.2.2. Since $T = \max\{X_1, X_2\}$, the CDF and PDF of T are

$$F_T(t) = P(T \le t) = F_1(t)F_2(t) = \begin{cases} F_{11}(t)F_{21}(t) & \text{if } t < \tau_1 \\ F_{12}(t-h_1)F_{22}(t-h_2) & \text{if } t \ge \tau_1, \end{cases}$$

and

$$f_T(t) = f_1(t)F_2(t) + f_2(t)F_1(t)$$
$$= \begin{cases} f_{11}(t)F_{21}(t) + f_{21}(t)F_{11}(t) & \text{if } t < \tau_1 \\ f_{12}(t-h_1)F_{22}(t-h_2) + f_{22}(t-h_1) + F_{12}(t-h_1) & \text{if } t \ge \tau_1, \end{cases}$$

respectively. Here h_j for $j = 1$ and 2 are the same as defined before, i.e., they satisfy Eq. (4.5). Hence, the joint PDF of T and Δ becomes

$$f_{T,\Delta}(t,j) = f_j(t)F_{j'}(t) = \begin{cases} f_{j1}(t)F_{j'1}(t) & \text{if } t < \tau_1 \\ f_{j2}(t-h_j)F_{j'2}(t-h_{j'}) & \text{if } t \ge \tau_1, \end{cases}$$

for $j, j' = 1, 2$ and $j \neq j'$. Therefore, based on the observations of the form (4.8) and using the same notations as in Section 4.2.2, the likelihood function of θ can be written as in Eq. (4.9).

Recently Han [46] considered the likelihood inference of the complementary risks model when the latent failure time distributions follow independent generalized exponential distributions with different shape and scale parameters. He obtained the MLEs of the unknown parameters, and proposed to use asymptotic distribution of the MLEs and the percentile bootstrap method to construct the associated confidence intervals. Extensive simulations have been performed to show the effectiveness of the proposed methods, and one real data set from an engineering case study regarding the reliability characteristics of a small/micro unmanned aerial vehicle has been analyzed for illustrative purposes.

Open Problems: There are several open problems associated with the step-stress model in the presence of complementary risks. Note that Han [46] has used the CEM assumption, but other step-stress models also will be of interest. The likelihood and Bayesian inference of other lifetime distributions need thorough investigations.

Miscellaneous topics

5

5.1 Introduction

In this chapter we consider different topics which are related to step-stress models but which we have not yet discussed. So far we have talked about different simple and multiple step-stress models when the stress changing time(s) is (are) prefixed. In this section we discuss some step-stress models when the stress changing time may be random. Xiong and Milliken [47] first considered the simple step-stress model when the stress changes immediately after a certain number of test units fails. Kundu and Balakrishnan [141] considered both the classical and Bayesian analysis of a simple step-stress model when the stress changing time is random and when the lifetime distributions of the experimental units follow exponential distributions at both the stress levels. The model can be extended to the multiple step-stress set-up in a straightforward manner. Ganguly and Kundu [133] provided the analysis for a competing risks model when the stress changing time is random, and when the lifetimes of the competing causes of failures follow exponential distributions. Recently, the results have been extended by Samanta and Kundu [142], to when the competing causes of failures follow Weibull distributions. In Section 5.2 we discuss different random stress changing time models and their associated inference.

One major aim in a step-stress life test (SSLT) is to increase the stress level to ensure early failures. Therefore, it is quite natural to assume that the mean lifetime decreases as the stress level increases. Balakrishnan et al. [143] first considered the ordered restricted inference of the step-stress model, for exponential distribution. It is observed that even for the simple exponential step-stress model the model parameters cannot be obtained in explicit form. One needs to use an isotonic regression technique to compute the order restricted MLEs. Since the MLEs cannot be obtained in explicit forms, the exact associated confidence intervals cannot be constructed very easily. Therefore, the Bayesian inference seems to be a reasonable choice. Recently Samanta et al. [144] considered the order restricted Bayesian inference for a simple step-stress model. The results can also be extended even for other lifetime distributions. We discuss different ordered restricted inference procedures in Section 5.3.

Kateri et al. [145] considered the analysis of a simple step stress model when the data have been collected from different experiments and where all the test units have been exposed to the same stress levels but may be with the different points of change of the stress levels. The authors provided a model that combines different experiments and provided a likelihood inference of the unknown parameters in case of complete sample and when the lifetimes of the experimental units follow exponential

Analysis of Step-Stress Models. http://dx.doi.org/10.1016/B978-0-12-809713-7.00005-3

distributions. The authors [146] further extended the results in case of Type-I censored sample. We discuss the meta-analysis approach for simple step-stress experiments in Section 5.4.

Designing an optimal step-stress experiment is an important problem. Extensive work has been done in designing an optimal simple or multiple step-stress experiment. The commonly used optimization criteria are (a) minimizing the sum of the asymptotic variances of the maximum likelihood estimates (MLEs), i.e., the sum of the diagonal elements of the inverse of the Fisher information matrix (A-optimality) and (b) maximizing the determinant of the Fisher information matrix (D-optimality). Miller and Nelson [147] first considered this optimality problem a simple step-stress model and when the lifetime of the experimental units follow exponential distribution at both the stress levels. Bai and Chung [108] and Bai et al. [148] extended the results of Miller and Nelson [147] for censored samples. Khamis and Higgins [117] and Khamis [149] considered the optimality issues for the multiple step-stress models in case of exponential distribution. Designing an optimal step-stress experiment has been considered for other lifetime distributions also; see for example Tang et al. [150] and Bai et al. [115] in this respect. Different optimality issues are discussed in Section 5.5. Finally, in Section 5.6 we provide several reference materials for further reading and future work.

5.2 Random stress changing time model

5.2.1 Experiment and model assumptions

Xiong and Milliken [47] first introduced the step-stress model when the stress changing time is random. The experiment can be described as follows. Suppose n items are used in a life testing experiment. Let $s_1, \ldots, s_k$ be k predetermined stress levels and $n_1, \ldots, n_{k-1}$ are $k-1$ prefixed nonnegative integers, such that $n_1 + \cdots + n_{k-1} < n$. At the beginning all the test units are subjected to the initial stress level s_1. As soon as the n_1th failure takes place, the stress level is changed to s_2 from s_1. Similarly, the stress level is increased from s_2 to s_3, as soon as the $(n_1 + n_2)$th failure takes place, and so on. Finally, when the $(n_1 + \cdots + n_{k-1})$th failure takes place, the stress level is changed from s_{k-1} to s_k, and the experiment continues till all the experimental units fail. The major advantage of this experiment is that it ensures n_j failures at the jth stress level for $j = 1, \ldots, k$, where $n_k = n - (n_1 + \cdots + n_{k-1})$. We provide the detailed analysis of the simple exponential step-stress model. The results can be easily extended for the multiple step-stress model, for details the readers are referred to Kundu and Balakrishnan [141].

5.2.2 Simple step-stress model: Exponential distribution

It is assumed that we have a simple step-stress model and the lifetime distributions at stress levels s_1 and s_2 are assumed to be exponential random variables with means θ_1 and θ_2, respectively. It is further assumed that we have Type-II censored data. Suppose we observe the following sample

$$\{t_{1:n} < \cdots < t_{n_1:n} < t_{n_1+1:n} < \cdots < t_{n_1+n_2:n}\}, \tag{5.1}$$

here $n_1 + n_2 = r < n$. Based on the assumptions of the cumulative exposure model (CEM), the joint probability density function (PDF) of $\{T_{1:n} < \cdots < T_{n_1+n_2:n}\}$ can be written as follows. The joint PDF of $\{T_{1:n} < \cdots < T_{n_1:n}\}$ is that of the first n_1 order statistics from a sample of size n from $\text{Exp}(\theta_1)$ and the conditional PDF of $\{T_{n_1+1:n} < \cdots < T_{n_1+n_2:n}\}$ given $\{T_{1:n} < \cdots < T_{n_1:n}\}$ is that of the smallest n_2 order statistics from a sample of size $n - n_1$ from $\text{Exp}(t_{n_1:n}, \theta_2)$. Therefore, the likelihood function of the observed data (Eq. 5.1) is

$$L(\theta_1, \theta_2) = \frac{c}{\theta_1^{n_1} \theta_2^{n_2}} e^{-\frac{T_1}{\theta_1}} e^{-\frac{T_2}{\theta_2}}, \tag{5.2}$$

where $c = n(n - 1) \cdots (n - r + 1)$, and

$$T_1 = \sum_{i=1}^{n_1} t_{i:n} + (n - n_1) t_{n_1:n} \tag{5.3}$$

$$T_2 = \sum_{i=n_1+1}^{n} (t_{i:n} - t_{n_1:n}) + (n - r)(t_{r:n} - t_{n_1:n}). \tag{5.4}$$

From Eq. (5.2), it is clear that (T_1, T_2) is a complete sufficient statistic. The MLEs of θ_1 and θ_2 can be obtained in explicit forms as follows:

$$\hat{\theta}_1 = \frac{T_1}{n_1} \quad \text{and} \quad \hat{\theta}_2 = \frac{T_2}{n_2}. \tag{5.5}$$

Kundu and Balakrishnan [141] obtained the joint PDFs of $\hat{\theta}_1$ and $\hat{\theta}_2$ and used them to construct the exact confidence intervals of θ_1 and θ_2. The following results are useful for further development.

Lemma 5.2.1. *Suppose* $Z_1, \ldots, Z_{n_1+n_2}$ *are* $(n_1 + n_2)$ *random variables with the joint PDF*

$$f_{Z_1,\ldots,Z_{n_1+n_2}}(t_1, \ldots, t_{n_1+n_2}) = \begin{cases} \frac{c}{\theta_1^{n_1} \theta_2^{n_2}} \times e^{-\frac{T_1}{\theta_1}} \times e^{-\frac{T_2}{\theta_2}} & \text{if } 0 < t_1 < \cdots < t_{n_1+n_2} \\ 0 & \text{otherwise,} \end{cases} \tag{5.6}$$

where c, T_1 *and* T_2 *are same as defined before. Consider the new set of random variables:*

$$Y_1 = \frac{Z_1}{\theta_1}, \ldots, Y_{n_1} = \frac{Z_{n_1}}{\theta_1}, Y_{n_1+1} = \frac{Z_{n_1+1} - Z_{n_1}}{\theta_2}, \ldots, Y_{n_1+n_2} = \frac{Z_{n_1+n_2} - Z_{n_1}}{\theta_2};$$

then the joint PDF of $Y_1, \ldots, Y_{n_1+n_2}$ *is*

$$f_{Y_1,\ldots,Y_{n_1+n_2}}(y_1,\ldots,y_{n_1+n_2}) = e^{-(\sum_{i=1}^{n_1} y_i+(n-n_1)y_{n_1})-(\sum_{i=n_1+1}^{n_1+n_2} y_i+(n-(n_1+n_2))y_{n_1+n_2}))},$$

for $0 < y_1 < \cdots < y_{n_1} < \infty$, $0 < y_{n_1+1} < \cdots < y_{n_1+n_2} < \infty$, *and zero otherwise.*

Proof. It can be obtained by a simple transformation technique. The details are omitted.   □

Lemma 5.2.2. *The joint moment generating function (MGF) of* $\frac{\hat{\theta}_1}{\theta_1}$ *and* $\frac{\hat{\theta}_2}{\theta_2}$ *is*

$$M(t,s) = Ee^{\left(\frac{t\hat{\theta}_1}{\theta_1} + \frac{s\hat{\theta}_2}{\theta_2}\right)} = \left(1 - \frac{t}{n_1}\right)^{-n_1}\left(1 - \frac{s}{n_2}\right)^{-n_2}; \quad |t| < n_1, \ |s| < n_2.$$

Proof. Note that

$$Ee^{\left(\frac{t\hat{\theta}_1}{\theta_1} + \frac{s\hat{\theta}_2}{\theta_2}\right)} = c\int_0^{\infty}\cdots\int_{y_{n_1}-1}^{\infty} e^{-(1-t)(\sum_{i=1}^{n_1} y_i+(n-n_1)y_{n_1})}dy_{n_1}\cdots dy_1$$

$$\times \int_0^{\infty}\cdots\int_{y_{n_1+n_2}-1}^{\infty} e^{-(1-s)(\sum_{i=n_1+1}^{n_1+n_2} y_i+(n-(n_1+n_2))y_{n_1+n_2})}$$

$$= \left(1 - \frac{t}{n_1}\right)^{-n_1}\left(1 - \frac{s}{n_2}\right)^{-n_2};$$

therefore, the result follows.   □

Using Lemmas 5.2.1 and 5.2.2, Kundu and Balakrishnan [141] obtained the following theorem regarding the joint distribution of the MLEs of the parameters:

Theorem 5.2.1. $\frac{\hat{\theta}_1}{\theta_1}$ *and* $\frac{\hat{\theta}_2}{\theta_2}$ *are distributed as Gamma*(n_1, n_1) *and Gamma*(n_2, n_2), *respectively, and they are independently distributed.*

Proof. It simply follows from Lemma 5.2.2.   □

Using Theorem 5.2.1, exact two-sided $100(1-\alpha)\%$ confidence intervals of θ_1 and θ_2 can be obtained as

$$\left(\frac{\hat{\theta}_1}{\gamma_{n_1,n_1}(1-\alpha/2)}, \frac{\hat{\theta}_1}{\gamma_{n_1,n_1}(\alpha/2)}\right) \quad \text{and} \quad \left(\frac{\hat{\theta}_2}{\gamma_{n_2,n_2}(1-\alpha/2)}, \frac{\hat{\theta}_2}{\gamma_{n_2,n_2}(\alpha/2)}\right),$$

$$(5.7)$$

respectively, where $\gamma_{k,k}(\delta)$ is the lower δth percentile point of a gamma(k,k) distribution. Alternatively, Eq. (5.7) can be written as

$$\left(\frac{2T_1}{\chi_{2n_1}^2(1-\alpha/2)}, \frac{2T_1}{\chi_{2n_1}^2(\alpha/2)}\right) \quad \text{and} \quad \left(\frac{2T_2}{\chi_{2n_2}^2(1-\alpha/2)}, \frac{2T_2}{\chi_{2n_2}^2(\alpha/2)}\right), \quad (5.8)$$

respectively, where $\chi_n^2(\alpha)$ is the lower αth percentile point of the χ_n^2 distribution. Kundu and Balakrishnan [141] proposed to use different bootstrap methods also to construct confidence intervals of θ_1 and θ_2. They have performed extensive simulation experiments to compare the performances of the different methods. It is observed that the exact confidence intervals and the bootstrap-t confidence intervals perform quite well in terms of the coverage percentages, and their average lengths are also very similar for both the parameters.

Kundu and Balakrishnan [141] provided the Bayesian inference of θ_1 and θ_2 based on independent gamma priors. The Bayes estimates (BEs) with respect to squared error loss function and the associated credible intervals can be obtained in explicit forms. It is observed that the performances of the BEs and the associated credible intervals are very similar with the MLEs and the exact confidence intervals, respectively, when the priors are noninformative. The authors have discussed different optimality related issues; interested readers are referred to the original paper of Kundu and Balakrishnan [141] in this respect.

Recently Ganguly and Kundu [133] considered the analysis of a simple step-stress model for exponentially distributed competing risks data and when the stress changing time is random. Based on the latent failure time model assumption of Cox [122], the authors obtained the MLEs of the unknown parameters and their exact distributions. Based on the exact distributions of the MLEs, confidence intervals of the unknown parameters can be obtained. The authors also provided several optimal test plans for this model.

Open Problem: Although Ganguly and Kundu [133] provided the classical inference of a simple step-stress model in the presence of the competing risks data, no work has been done regarding the Bayesian inference of this model. It will be interesting to develop the Bayesian inference of this model and to compare the results with the classical ones.

In the next section we discuss the inferential issues of the Weibull models when the stress changing time is random and the model satisfies the tampered failure rate model (TFRM) assumptions.

5.2.3 Simple step-stress model: Weibull distribution

In this case we have the same experimental setup as in the previous section and the observed sample is of the form (5.1). It is further assumed that the PDF of the lifetime of an item at the stress level s_i is given by

$$f(t; \alpha, \theta_i) = \frac{\alpha}{\theta_i} t^{\alpha-1} e^{-\frac{t^{\alpha}}{\theta_i}} \mathbb{1}_{[0, \infty)}(t), \tag{5.9}$$

for $i = 1$ and 2. Since it is assumed that the model satisfies the TFRM assumption, and if T denotes the lifetime under this random step-stress model, then the conditional cumulative distribution function (CDF) of T given that $t_{n_1:n} = c$ is given by

$$F(t) = \left(1 - e^{-\frac{t^{\alpha}}{\theta_1}}\right) \mathbb{1}_{[0, c)}(t) + \left(1 - e^{-\frac{t^{\alpha} - c^{\alpha}}{\theta_2} - \frac{c^{\alpha}}{\theta_1}}\right) \mathbb{1}_{(c, \infty)}(t). \tag{5.10}$$

Observe that the distribution of $T_{1:n} < \cdots < T_{n_1:n}$ is same as that of the first n_1 order statistics from i.i.d. random variables of sample size n from the lifetime distribution under the stress level s_1, and for given $T_{1:n} < \cdots < T_{n_1:n}$, the distribution of $T_{n_1+1:n} < \cdots < T_{n_1+n_2:n}$ is the same as that of the first n_2 order statistics from an i.i.d. random sample of size $n - n_1$ from the following PDF:

$$g(t; \alpha, \theta_2) = \begin{cases} 0 & \text{if } t < t_{n_1:n} \\ \frac{\alpha}{\theta_2} t^{\alpha-1} e^{-\frac{t^\alpha - t_{n_1:n}^\alpha}{\theta_2}} & \text{if } t \geq t_{n_1:n}. \end{cases} \qquad (5.11)$$

Let us use the following notations:

$$D_1(\alpha) = \sum_{i=1}^{r} t_{i:n}^\alpha + (n - n_1)\, t_{n_1:n}^\alpha$$

$$D_2(\alpha) = \sum_{i=n_1+1}^{n_1+n_2} (t_{i:n}^\alpha - t_{n_1:n}^\alpha) + (n - n_1 - n_2)(t_{n_1+n_2:n}^\alpha - t_{n_1:n}^\alpha).$$

The joint PDF of $T_{1:n} < \cdots < T_{n_1:n}$, when $c_1 = \frac{n!}{(n-n_1)!}$, becomes

$$f_{T_{1:n},\ldots,T_{n_1:n}}(t_{1:n}, \ldots, t_{n_1:n}) = c_1 \left(\frac{\alpha}{\theta_1}\right)^{n_1} \left(\prod_{i=1}^{r} t_{i:n}\right)^{\alpha-1} e^{-\frac{D_1(\alpha)}{\theta_1}},$$

for $t_{1:n} < \cdots < t_{n_1:n}$ and 0, otherwise. The joint PDF of $T_{n_1+1:n} < \cdots < T_{n_1+n_2:n}$ given $T_{1:n} < \cdots < T_{n_1:n}$ is

$$f_{T_{n_1+1:n},\ldots,T_{n_1+n_2:n}|T_{1:n},\ldots,T_{n_1:n}}(t_{n_1+1:n}, \ldots, t_{n_1+n_2:n})$$

$$= c_2 \left(\frac{\alpha}{\theta_2}\right)^{n_2} \left(\prod_{i=n_1+1}^{s} t_{i:n}\right)^{\alpha-1} e^{-\frac{D_2(\alpha)}{\theta_2}},$$

for $t_{1:n} < \cdots < t_{n_1:n} < t_{n_1+1:n} < \cdots < t_{n_1+n_2:n}$, and 0, otherwise. Here $c_2 = \frac{(n-r)!}{(n-n_1-n_2)!}$. Hence, the joint PDF of $T_{1:n} < \cdots < T_{n_1+n_2:n}$ is given by

$$f_{T_{1:n},\ldots,T_{n_1+n_2:n}}(t_{1:n}, \ldots, t_{n_1+n_2:n}) = c_3 \times \frac{\alpha^{n_1+n_2}}{\theta_1^{n_1} \theta_2^{n_2}} \left(\prod_{i=1}^{n_1+n_2} t_{i:n}\right)^{\alpha-1} e^{-\frac{D_1(\alpha)}{\theta_1} - \frac{D_2(\alpha)}{\theta_2}},$$

$$(5.12)$$

for $t_{1:n} < \cdots < t_{n_1+n_2:n}$ and 0, otherwise, with $c_3 = \frac{n!}{(n-n_1-n_2)!}$.

Now we would like to obtain the MLEs of the unknown parameters. From the joint PDF of $T_{1:n} < \cdots < T_{n_1+n_2:n}$ as in Eq. (5.12), the log-likelihood function of α, θ_1 and θ_2 can be written as

$$l(\alpha, \theta_1, \theta_2) = c + (n_1 + n_2)\ln\alpha - n_1\ln\theta_1 - n_2\ln\theta_2 + (\alpha-1)\sum_{i=1}^{n_1+n_2}\ln t_{i:n} - \frac{D_1(\alpha)}{\theta_1} - \frac{D_2(\alpha)}{\theta_2},$$

$$(5.13)$$

where $c = \sum_{i=1}^{s}\ln(n - i + 1)$. For fixed $\alpha > 0$, Eq. (5.13) can be maximized with respect to θ_1 and θ_2, and they can be obtained as

$$\hat{\theta}_1(\alpha) = \frac{D_1(\alpha)}{n_2} \quad \text{and} \quad \hat{\theta}_2(\alpha) = \frac{D_2(\alpha)}{n_2}, \qquad (5.14)$$

respectively. Now replacing θ_1 and θ_2 by $\hat{\theta}_1(\alpha)$ and $\hat{\theta}_2(\alpha)$, respectively, in Eq. (5.13), we have the profile log-likelihood function of α. The MLE of α can be found by maximizing the profile log-likelihood of α. Note that the maximization of the profile log-likelihood function of α is same as the maximization of the following function:

$$l^*(\alpha) = (n_1 + n_2)\ln\alpha - n_1\ln D_1(\alpha) - n_2\ln D_2(\alpha) + (\alpha - 1)\sum_{i=1}^{n_1+n_2}\ln t_{i:n}. \quad (5.15)$$

However, the MLE of α cannot be obtained in explicit form. One needs to use some numerical methods to compute the MLE. Once the MLE of α is obtained the MLEs of θ_1 and θ_2 can be obtained from Eq. (5.14) by replacing α with the MLE of α.

Now we will show that Eq. (5.15) possesses a unique maximum in $(0, \infty)$ if $n_1 > 1$ and $n_2 > 1$. We need the following lemma for that purpose.

Lemma 5.2.3. *The function*

$$\Upsilon(\alpha) = \frac{\sum_{j=k_1+1}^{k_2} x_j^\alpha\ln x_j - (n - k_1)t_1^\alpha\ln t_1 + (n - k_2)t_2^\alpha\ln t_2}{\sum_{j=k_1+1}^{k_2} x_j^\alpha - (n - k_1)t_1^\alpha + (n - k_2)t_2^\alpha} - \frac{1}{\alpha}$$

is a strictly increasing function in $\alpha > 0$, where k_1 and k_2 are integers, $k_1 < k_2$, and $0 \le t_1 \le x_{k_1+1} \le \cdots \le x_{k_2} \le t_2$.

Proof. See Lemma 2 of Wang and Fei [116]. □

Note that $l_\alpha^* = 0$ can be written as

$$\frac{n_1}{n_1 + n_2} \times \left(\frac{D_1'(\alpha)}{D_1(\alpha)} - \frac{1}{\alpha}\right) + \frac{n_2}{n_1 + n_2} \times \left(\frac{D_2'(\alpha)}{D_2(\alpha)} - \frac{1}{\alpha}\right) = \frac{1}{n_1 + n_2}\sum_{i=1}^{n_1+n_2}\ln t_{i:n},$$

$$(5.16)$$

Using Lemma 5.2.3, it easily follows that both the functions

$$g_1(\alpha) = \left(\frac{D_1'(\alpha)}{D_1(\alpha)} - \frac{1}{\alpha}\right) \quad \text{and} \quad g_2(\alpha) = \left(\frac{D_2'(\alpha)}{D_2(\alpha)} - \frac{1}{\alpha}\right)$$

are strictly increasing functions in $\alpha > 0$. Further, it can be shown that

$$\lim_{\alpha \to 0+} g_1(\alpha) = -\infty, \quad \lim_{\alpha \to \infty} g_1(\alpha) = \ln t_{n_1:n}$$

and

$$\lim_{\alpha \to 0+} g_2(\alpha) = -\infty, \quad \lim_{\alpha \to \infty} g_2(\alpha) = \ln t_{n_1+n_2:n}.$$

Hence, for $n_1 > 1$ and $n_2 > 1$, Eq. (5.16) has a unique solution.

Since the MLEs cannot be obtained in explicit forms, finding the exact distributions of the MLEs is difficult. One may use the bootstrap method to construct confidence intervals of the unknown parameters. Recently, Zhang and Shi [151] considered the same problem under Type-II adaptive progressive hybrid censoring schemes. They obtained the MLEs of the unknown parameters and provided the associated confidence intervals.

5.2.4 Multiple step stress model

Balakrishnan et al. [152] also considered the same problem based on a sequential order statistics approach. It is assumed that at the stress level j, if the underlying CDF of the experimental units is assumed to be $F_j(\cdot)$, then $F_j(\cdot)$ satisfies the following condition:

$$F_j(t) = 1 - \{1 - F(t)\}^{1/\theta_j}, \tag{5.17}$$

for $j = 1, \ldots, k$. Here $\theta_1 > 0, \ldots, \theta_k > 0$ are the unknown parameters and $F(\cdot)$ is a baseline absolutely continuous distribution function which is assumed to be known. It may be noted that this model does not satisfy the CEM assumptions in general, except when the baseline distribution is exponential. In case of the exponential distribution the CEM and the preceding model match.

For a given baseline distribution function $F(t)$, the joint PDF of the observed failure times can be obtained using the sequential order statistics approach of Kamps [153]. The unique MLEs of $\theta_1, \ldots, \theta_k$ can be obtained as

$$\hat{\theta}_1 = -\frac{1}{n_1} \sum_{i=1}^{n_1} (n - i + 1) \left(\ln \bar{F}(t_{i:n}) - \ln \bar{F}(t_{i-1:n}) \right)$$

$$\hat{\theta}_j = -\frac{1}{n_j} \sum_{i=n_1+\cdots+n_{j-1}+1}^{n_1+\cdots+n_j} (n - i + 1) \left(\ln \bar{F}(t_{i:n}) - \ln \bar{F}(t_{i-1:n}) \right); \quad j = 2, \ldots, k.$$

Here $\bar{F}(t_{0:n}) = 1$. It can be shown that $\hat{\theta}_1, \ldots, \hat{\theta}_k$ are unbiased estimators of $\theta_1, \ldots, \theta_k$, respectively, and they are independently distributed. Based on the exact distributions of $\hat{\theta}_1, \ldots, \hat{\theta}_k$, exact confidence intervals of the unknown parameters can be constructed.

Open Problem: Balakrishnan et al. [152] considered the problem based on the assumption that the baseline distribution function $F(t)$ is completely known. It will be important to consider the model when $F(t)$ also has some unknown parameters, and develop a statistical inference of the model parameters. The analysis can also be extended when the baseline distribution $F(\cdot)$ is completely unknown.

5.3 Order restricted inference

5.3.1 Classical inference

Balakrishnan et al. [143] first considered the ordered restricted inference of the unknown parameters for the Type-I and Type-II censored data using the CEM assumptions. We will describe the procedure for Type-I censored data; the other can be obtained along the same line. Let us recall that n identical units are placed on a life testing experiment at the initial stress level s_1, and at the prefixed time points $\tau_1 < \cdots < \tau_{k-1}$, and the stress level increases to $s_2, \ldots, s_k$, respectively. The experiment continues till the time point $\tau_k > \tau_{k-1}$.

It is assumed that the lifetime distribution of the experimental units follows an exponential distribution. At the stress level s_i, the mean lifetime of the experimental units is θ_i, for $i = 1, \ldots, k$. It is further assumed that

$$\theta_1 \geq \theta_2 \geq \cdots \geq \theta_k. \tag{5.18}$$

Although this order restriction is very intuitive, in developing the inference of the unknown parameters, this restriction has been ignored in most of the studies, mainly because of the analytical intractability.

Based on the CEM assumptions as in Section 2.8, the PDF of the lifetime of a test unit from the preceding experiment can be written as

$$f(t) = \begin{cases} 0 & \text{if } t < 0 \\ \frac{1}{\theta_1} \exp\left(-\frac{t}{\theta_1}\right) & \text{if } 0 \leq t < \tau_1 \\ \frac{1}{\theta_m} \exp\left(-\frac{t-\tau_{m-1}}{\theta_m} - \sum_{j=1}^{m-1} \frac{\tau_j-\tau_{j-1}}{\theta_j}\right) & \text{if } \tau_{m-1} \leq t < \tau_m; \quad m = 2, \ldots, k-1, \\ \frac{1}{\theta_k} \exp\left(-\frac{t-\tau_{k-1}}{\theta_k} - \sum_{j=1}^{k-1} \frac{\tau_j-\tau_{j-1}}{\theta_j}\right) & \text{if } \tau_{k-1} \leq t < \infty; \end{cases} \tag{5.19}$$

see for example Section 2.8. We use the following notations for further development.

Let $T_1, \ldots, T_n$ be the failure times of the n experimental units. We denote $\boldsymbol{T} = (T_{1:n}, \ldots, T_{\bar{n}:n})$ as the vector of ordered failure times until the termination of the experiment and $\boldsymbol{t} = (t_{1:n}, \ldots, t_{\bar{n}:n})$ as a realization of $\boldsymbol{T}$. Here $\bar{n}$ denotes the total number of observed failures till τ_k and it is a random variable. Let

$$N_i(\boldsymbol{T}) = \#\{T_1, \ldots, T_n | \tau_{i-1} \leq T_j < \tau_i\}; \quad i = 1, \ldots, k.$$

Therefore, $N_i(\boldsymbol{T})$ denotes the random number of failures that occurs at the stress level s_i for $i = 1, \ldots, k$. Let

$$\bar{N}_i(\boldsymbol{T}) = N_1(\boldsymbol{T}) + \cdots + N_i(\boldsymbol{T}); \quad i = 1, \ldots, k,$$

$n_i(t)$ and $\bar{n}_i(t)$ denote the realizations of $N_i(\boldsymbol{T})$ and $\bar{N}_i(\boldsymbol{T})$, respectively. We further denote $I = \{1, \ldots, k\}$,

$$\tilde{I} = \{i \in I | N_i(T) \geq 1\}$$

and the elements of $\tilde{I}$ as $i_1, i_2, \ldots, i_{|\tilde{I}|}$.

Then based on the observed vector t, using the preceding notations, the likelihood function can be written as

$$L(\theta_1, \ldots, \theta_k) = \left(\prod_{i=1}^{k} \frac{1}{\theta_i^{n_i(t)}}\right) \exp\left(-\sum_{i=1}^{k} \frac{d_i}{\theta_i}\right), \qquad (5.20)$$

where $\bar{n}_0(t) = 0$, $\tau_0 = 0$, and

$$d_i = \sum_{j=\bar{n}_{i-1}(t)+1}^{\bar{n}_i(t)} (t_{j:n} - \tau_{i-1}) + (n - \bar{n}_i(t))(\tau_i - \tau_{i-1}); \quad i = 1, \ldots, k.$$

Therefore, the MLEs of the unknown parameters can be obtained by maximizing Eq. (5.20) with respect to $\theta_1, \ldots, \theta_k$, under the order restriction (5.18). It is immediate that without any restriction on the parameters, the MLE of θ_i exists only if $i \in \tilde{I}$, and in this case the MLE is given by

$$\hat{\theta}_i = \frac{d_i}{n_i(t)}; \quad i \in \tilde{I}. \qquad (5.21)$$

If $i \in \tilde{I}^c$, the MLE of θ_i does not exist. It should be made clear that the maximization of the likelihood function is performed only with respect to those θ_i that affect the likelihood function (5.20). Clearly, if for any $i \notin \tilde{I}$, d_i becomes zero, then Eq. (5.20) does not depend on θ_i.

Balakrishnan et al. [143] showed, using the isotonic regression technique, that the MLE $\breve{\theta}_m$ of θ_m under the order restriction (5.18) is given by

$$\breve{\theta}_m = \min_{\substack{s \leq m \\ s \in \tilde{I}}} \max_{\substack{t \geq m \\ t \in \tilde{I}}} \frac{\sum_{j=s, j \in \tilde{I}}^{t} n_j(t)\hat{\theta}_j}{\sum_{j=s, j \in \tilde{I}}^{t} n_j(t)}; \quad m \in \tilde{I}. \qquad (5.22)$$

Moreover, if we denote $\tilde{I} = \{i_1, i_2, \ldots, i_{m*}\}$, where $i_1 < \cdots < i_{m*}$, then the MLE $\breve{\theta}_m$ of θ_m for $i_1 \leq m \leq i_{m*}$ is given by

$$\breve{\theta}_m = \breve{\theta}_{i(m)}; \quad \text{where} \quad i(m) = \max\{i \in \tilde{I} | i \leq m\}. \qquad (5.23)$$

The authors obtained the confidence intervals of the unknown parameters based on the asymptotic distribution of the MLEs.

Recently, Samanta et al. [154] considered the same problem and provided a simple algorithm to compute the MLEs of the unknown parameters. Consider the following transformation of the parameters from $(\theta_1, \ldots, \theta_k)$ to $(\theta_1, \beta_1, \ldots, \beta_{k-1})$, where

$$\theta_2 = \beta_1 \theta_1$$
$$\theta_3 = \beta_2 \theta_2 = \beta_1 \beta_2 \theta_1$$
$$\vdots$$
$$\theta_k = \beta_{k-1} \theta_{k-2} = \beta_1 \cdots \beta_{k-1} \theta_1. \tag{5.24}$$

Clearly, the transformation (5.24) is a one-to-one transformation and maximizing the likelihood function (5.20) under the constraints (5.18) is equivalent to maximizing the likelihood function (5.20) under the constraints

$$0 < \theta_1 < \infty, \quad 0 \le \beta_j \le 1, \quad \text{for} \quad j = 1, \ldots, k-1. \tag{5.25}$$

Therefore, the MLEs of $\theta_1, \ldots, \theta_k$ under the constraints (5.18) can be obtained by first finding the MLEs of $\theta_1, \beta_1, \ldots, \beta_{k-1}$ under the constraints (5.25) and then using the one-to-one transformation between $(\theta_1, \ldots, \theta_k)$ and $(\theta_1, \beta_1, \ldots, \beta_{k-1})$ as given in Eq. (5.24). The MLEs of $\theta_1, \beta_1, \ldots, \beta_{k-1}$ under the constraints (5.25) can be obtained as follows. From Eq. (5.20) it can be easily seen that the log-likelihood function of the observation $t = (t_{1:n}, \ldots, t_{\tilde{n}:n})$ based on the parameters $\theta_1, \beta_1, \ldots, \beta_{k-1}$ can be written as

$$l(\theta_1, \beta_1, \ldots, \beta_{k-1}) = -n \ln \theta_1 - \sum_{j=1}^{k-1} \bar{n}_{j+1} \ln \theta_j - \frac{1}{\theta_1} \left[\sum_{j=1}^{k} \frac{d_j}{\prod_{i=1}^{j-1} \beta_i} \right]. \tag{5.26}$$

Hence, the MLEs can be obtained by maximizing Eq. (5.26) with respect to $\theta_1, \beta_1, \ldots, \beta_{k-1}$ under the constraints (5.25). First let us assume $\tilde{I} = I$. In this case, it can be easily seen that the function (5.26) has a unique global maximum at the point $(\theta_1^*, \beta_1^*, \ldots, \beta_{k-1}^*)$, where

$$\theta_1^* = \frac{d_1}{n_1}, \quad \beta_1^* = \frac{d_2 n_1}{n_2 d_1}, \ldots, \quad \beta_{k-1}^* = \frac{d_k n_{k-1}}{n_k d_{k-1}}. \tag{5.27}$$

Note that $\theta_1^*, \beta_1^*, \ldots, \beta_{k-1}^*$ may not be MLEs of $\theta_1, \beta_1, \ldots, \beta_{k-1}$, respectively, under the constraints (5.25). Samanta et al. [154] suggested the following algorithm to compute the MLEs of $\theta_1, \beta_1, \ldots, \beta_{k-1}$ under the constraints (5.25).

Algorithm 5.1
Step 1: For the given data $t = (t_{1:n}, \ldots, t_{\tilde{n}:n})$ obtain $\theta_1^*, \beta_1^*, \ldots, \beta_{k-1}^*$ as given in Eq. (5.27). If $\beta_j^* \le 1$, for all $j = 1, \ldots, k-1$, then $\theta_1^*, \beta_1^*, \ldots, \beta_{k-1}^*$ are the MLEs of $\theta_1, \beta_1, \ldots, \beta_{k-1}$, respectively, under the constraints (5.25).
Step 2: If one or some of the $\beta_j^* > 1$, then replace them by 1, in the log-likelihood function (5.26) and reestimate the remaining parameters by maximizing the modified log-likelihood function.
Step 3: If all the new estimates of β_j's are less than or equal to one, then they are the MLEs, and the corresponding estimate of θ_1 is the MLE of θ_1.
Step 4: If all the new estimates of β_j's are not less than or equal to one, replace them by one, and continue the process unless all the estimates of β_j's are less than or equal to one.

Note that in each step the estimates of θ_1 and $\beta_1, \ldots, \beta_{k-1}$ can be obtained in explicit forms, and Algorithm 5.1 stops in a finite number of steps. Moreover, it has been shown by Samanta et al. [154] that Algorithm 5.1 provides the MLEs of $\theta_1, \beta_1, \ldots, \beta_{k-1}$ under the constraints (5.25).

Further it has been shown that if $\tilde{I} = \{i_1, i_2, \ldots, i_{m^*}\}$, where $1 \leq i_1 < i_2 < \cdots < i_{m^*} \leq k$, then the MLEs of θ_m, for $i_1 \leq m \leq i_{m^*}$, can be obtained along the same manner. Since the MLEs cannot be obtained in explicit forms, finding the exact distribution of the MLEs is quite difficult. Due to this reason, the authors have used the observed Fisher information matrix and the bootstrap method to construct approximate confidence intervals of the unknown parameters.

5.3.2 Bayesian inference

In the previous subsection it has been observed that, although the MLEs of the unknown parameters under the order restriction can be obtained, finding the exact distribution of the MLEs is quite difficult. Therefore, the Bayesian inference seems to be a reasonable choice in this case. Samanta et al. [154] considered the Bayesian inference of $\theta_1, \ldots, \theta_k$ under the order restriction (5.18) based on the same reparameterization (5.24).

Let us assume that θ_1 has a prior distribution $\pi_0(\theta_1)$ and β_i has a prior distribution $\pi_i(\beta_i)$, for $i = 1, \ldots, k - 1$. Then the following assumptions have been made on $\pi_0(\theta_1), \pi_1(\beta_1), \ldots, \pi_{k-1}(\beta_{k-1})$:

$$\theta_1 \sim \text{IGamma}(a_0, b_0) \quad \text{and} \quad \beta_i \sim \text{Beta}(a_i, b_i); \quad i = 1, \ldots, k-1, \tag{5.28}$$

and $\theta_1, \beta_1, \ldots, \beta_{k-1}$ are all independently distributed. The joint prior distribution of $\theta_1, \beta_1, \ldots, \beta_{k-1}$ can be written as

$$\pi(\theta_1, \beta_1, \ldots, \beta_{k-1}) \propto e^{-\frac{b_0}{\theta_1}} \left(\frac{1}{\theta_1}\right)^{a_0+1} \prod_{j=1}^{k-1} \beta_j^{a_j-1}(1 - \beta_j)^{b_j-1}, \tag{5.29}$$

for $\theta_1 > 0$ and $0 \leq \beta_1, \ldots, \beta_{k-1} \leq 1$. The posterior distribution of $\theta_1, \beta_1, \ldots, \beta_{k-1}$ can be written as

$$l(\theta_1, \beta_1, \ldots, \beta_{k-1}|t) \propto l_1(\beta_1|t), \ldots, l_{k-1}(\beta_{k-1}|t)l_k(\theta_1|\beta_1, \ldots, \beta_{k-1}, t)$$
$$\times w(\theta_1, \beta_1, \ldots, b_{k-1}, t). \tag{5.30}$$

Here, $l_i(\beta_i) = 1$ for $i = 1, \ldots, k - 1$, $l_k((\theta_1|\beta_1, \ldots, \beta_{k-1}, Data)$ is the PDF of an IGamma$(n + a_0 + 1, b_0 + A(\beta_1, \ldots, \beta_{k-1}, t)$, and

$$w(\theta_1, \beta_1, \ldots, b_{k-1}, t) = \frac{\prod_{i=1}^{k-1} \beta_i^{a_i - \bar{n}_{i+1} - 1} (1 - \beta_i)^{b_i - 1}}{[b_0 + A(\beta_1, \ldots, \beta_{k-1}, t)]^{n+a_0}},$$

$$A(\beta_1, \ldots, \beta_{k-1}, t) = \sum_{i=1}^{k} \frac{d_i}{\prod_{j=0}^{i-1} \beta_j}, \qquad \beta_0 = 1.$$

Therefore, under the squared error loss function the BE of any parametric function of $\theta_1, \beta_1, \ldots, \beta_{k-1}$, say $g(\theta_1, \beta_1, \ldots, \beta_{k-1})$, is the posterior expectation of $g(\theta_1, \beta_1, \ldots, \beta_{k-1})$ provided it exists and it is given by

$$
\begin{aligned}
\hat{g}_B &= \hat{g}_B(\theta_1, \beta_1, \ldots, \beta_{k-1}) \\
&= E_{\theta_1, \beta_1, \ldots, \beta_{k-1} | t} (g(\theta_1, \beta_1, \ldots, \beta_{k-1})) \\
&= \int_0^1 \cdots \int_0^1 \int_0^\infty g(\theta_1, \beta_1, \ldots, \beta_{k-1}) l(\theta_1, \beta_1, \ldots, \beta_{k-1} | t) d\theta_1 d\beta_1 \cdots d\beta_{k-1}.
\end{aligned}
$$
(5.31)

In general it is not possible to obtain Eq. (5.31) in explicit form. Samanta et al. [154] proposed the following algorithm to compute the BE and the associated credible interval.

Algorithm 5.2
Step 1: Generate $\beta_{11}, \ldots, \beta_{(k-1)1}$ from a U(0, 1) distribution.
Step 2: For a given set of $\{\beta_{11}, \ldots, \beta_{(k-1)1}\}$, generate θ_{11} from IGamma($n + a_0, b_0 + A(\beta_{11}, \ldots, \beta_{(k-1)1})$).
Step 3: Repeat Step 1 and Step 2 M times to obtain $(\beta_{11}, \ldots, \beta_{(k-1)1}, \theta_{11}), \ldots, (\beta_{1M}, \ldots, \beta_{(k-1)M}, \theta_{1M})$.
Step 4: Compute $g_i = g(\theta_{1i}, \beta_{1i}, \ldots, \beta_{(k-1)i})$ for $i = 1, \ldots, M$.
Step 5: Calculate the weights $w_i = \dfrac{w(\theta_i, \beta_{1i}, \ldots, \beta_{(k-1)i})}{\sum_{i=1}^{M} w(\theta_{1i}, \beta_{1i}, \ldots, \beta_{(k-1)i})}$.
Step 6: Obtain the BE of $g(\theta_1, \ldots, \theta_{k-1})$ under the squared error loss function as

$$\hat{g}(\theta_1, \beta_1, \ldots, \beta_{k-1}) = \sum_{i=1}^{M} w_i g_i.$$

Step 7: To construct a $100(1 - \gamma)\%$, for $0 < \gamma < 1$, credible interval of $g(\theta_1, \ldots, \theta_{k-1})$, first order $g_j; j = 1, \ldots, M$, say, $g_{(1)} < \cdots < g_{(M)}$, and arrange w_j accordingly to get $w_{[1]}, \ldots, w_{[M]}$. Clearly, $w_{[1]}, \ldots, w_{[M]}$ may not be ordered.
Step 8: A $100(1 - \gamma)\%$ credible interval of $g(\theta_1, \ldots, \theta_{k-1})$ can be obtained as $(g_{(j_1)}, g_{(j_2)})$, where $j_1, j_2 \in \{1, \ldots, M\}, j_1 < j_2$ and they satisfy

$$\sum_{i=j_1}^{j_2} w_{[i]} \leq 1 - \gamma < \sum_{i=j_1}^{j_2+1} w_{[i]}.$$
(5.32)

The $100(1 - \gamma)\%$ HPD credible interval of $g(\theta_1, \ldots, \theta_{k-1})$ can be obtained as $(g_{(j_1^*)}, g_{(j_2^*)})$, where $j_1^*, j_2^* \in \{1, \ldots, M\}, j_1^* < j_2^*$ and they satisfy Eq. (5.32). Moreover,

$$g_{(j_2^*)} - g_{(j_1^*)} \leq g_{(j_2)} - g_{(j_1)},$$

for all j_1, j_2 satisfying Eq. (5.32).

Extensive simulations have been performed by Samanta et al. [154] to observe the performances of the MLEs and the Bayes estimators. The performances of the MLEs and the BEs with respect to noninformative priors are very similar in terms of the biases and MSEs. The performances of the height posterior density (HPD) credible intervals and the confidence intervals obtained by the bootstrap method or based on the Fisher information matrix are very similar in terms of coverage percentages. The average lengths of the HPD credible intervals are slightly smaller than those of the confidence intervals mentioned previously.

5.4 Meta-analysis approach

We have discussed different step-stress models that are based on a single experiment. Kateri et al. [155] first considered a meta-analysis approach for different step-stress experiments. The authors provided a method to combine data obtained from different step-stress experiments. Suppose step-stress experiments were conducted in more than one places and at different stress levels. Moreover, the sample sizes may not be the same, and naturally they may provide a different number of observed failures. One natural question is how to combine all these different data sets and provide inference of the unknown parameters. Therefore, it is a natural generalization from the one-sample problem to the multisample problem.

The problem can be stated as follows. Suppose g independent samples of sizes $n_1, \ldots, n_g$ of the same population have been placed on a life testing experiment at the initial stress level s_1. The stress level is increased to $s_2, \ldots, s_k$ at the prefixed time points $\tau_{1j} < \cdots < \tau_{k-1,j}$, respectively, in the jth sample, $j = 1, \ldots, g$. Now based on the observed failures at g different groups how can we draw inferences of the unknown parameters.

Kateri et al. [155] considered the simple step-stress model and it is assumed that the lifetime of the experimental units follow an exponential distribution with parameters θ_1 and θ_2 at two different stress levels for all the g samples. It is also assumed that the experiment continues until a prespecified number of failures $r_j, j = 1, 2, \ldots, g$ is observed in the jth sample. Now based on the CEM assumptions $\hat{\theta}_1$ and $\hat{\theta}_2$, the conditional MLEs of θ_1 and θ_2, respectively, can be obtained in explicit forms. The exact distributions $\hat{\theta}_1$ and $\hat{\theta}_2$ have been obtained using the conditional moment generating function (CMGF). As observed for one sample case, the conditional distribution of $\hat{\theta}_1$ is a generalized mixture of translated gamma distributions and the conditional distribution of $\hat{\theta}_2$ is a pure mixture of gamma distributions.

Open Problems: The authors have provided the inference procedure only for the exponential distribution, but the results can be extended for other lifetime distributions. It would be interesting to develop Bayesian inference in this case.

5.5 Optimal design of SSLTs

We have discussed the estimation procedures of the unknown parameters for different step-stress models under various censoring schemes. Here we discuss a very important issue of designing an *optimum* step-stress experiment. Consider a step-stress life test with the k stress levels $s_1, s_2, \ldots, s_k$, and the stress levels are changed at the times $\tau_1 < \tau_2 < \cdots < \tau_{k-1}$. In this section we discuss the *optimum* choice of the stress changing times $\tau_1, \tau_2, \ldots, \tau_{k-1}$.

Suppose that the lifetime of an experimental unit follows an exponential distribution with mean θ_i at the stress level s_i, $i = 1, 2, \ldots, k$. Let s_L and s_H be the lowest and highest allowable stress levels, respectively. Note that s_L is at least as severe as the use or normal stress level. Usually s_L is chosen as low as possible, subject to the constraint that the test ends by the desired time. Similarly, s_H is chosen as high as possible, subject to the condition that the mechanism of failure does not change from that of the normal or used stress level. In general, a link function is used to describe the relationship between the stress level and the corresponding lifetime distribution. Some popular link functions that have been used by several researchers are linear, log-linear, and log-quadratic functions. Let us assume that the link function is given by

$$\theta_i = \psi(x_i), \quad i = 1, 2, \ldots, k,$$

where $x_i = \frac{s_i - s_L}{s_H - s_L}$ and $\psi(\cdot)$ involves several parameters. For example if we consider $\psi(x_i) = e^{\alpha + \beta x_i}$, then the link function is given by

$$\ln \theta_i = \alpha + \beta x_i,$$

where α and β are unknown parameters.

Now we will discuss some popular optimality criteria, most of which are defined using the Fisher information matrix. Let I be the Fisher information matrix of the unknown parameters.

A-Optimal Criteria: This optimal criterion is based on the sum of asymptotic variances of the MLEs of the unknown parameters, i.e., it is the sum of diagonal elements of the inverse of the Fisher information matrix I. Thus the objective function of the A-optimal criterion is given by

$$\phi_A(\tau_1, \tau_2, \ldots, \tau_{k-1}) = \text{Tr}(I),$$

where Tr stands for the trace of a matrix. A-optimal $\tau_1, \tau_2, \ldots, \tau_{k-1}$ can be found by minimizing $\phi_A(\cdot, \cdot, \ldots, \cdot)$ over the ordered region $\tau_1 < \tau_2 < \cdots < \tau_{k-1}$.

C-Optimal Criteria: In a life testing experiment, the researchers often want to estimate the mean lifetime at the used condition with a maximum precision. Motivated by this argument, the next objective function is

$$\phi_c(\tau_1, \tau_2, \ldots, \tau_{k-1}) = n \, \mathrm{AVar}\left(\ln \hat{\theta}_0\right),$$

where AVar stands for the asymptotic variance that can be computed using the Fisher information matrix. Here θ_0 is the mean lifetime at the normal stress level. C-optimal $\tau_1, \tau_2, \ldots, \tau_{k-1}$ can be found by minimizing $\phi_c(\cdot, \cdot, \ldots, \cdot)$ over the ordered region $\tau_1 < \tau_2 < \cdots < \tau_{k-1}$.

D-Optimal Criteria: This optimality criterion is based on the determinant of the Fisher information matrix. Note that the volume of the joint confidence region of the unknown parameters is directly proportional to the positive square root of the determinant of the inverse of the Fisher information matrix I at a fixed confidence level. In other words a large value of $|I|$ implies a small joint confidence ellipsoid of the unknown parameters. Motivated by this argument, the objective function for D-optimality is given by

$$\phi_D(\tau_1, \tau_2, \ldots, \tau_{k-1}) = |I|^{\frac{1}{2}}.$$

The D-optimal $\tau_1, \tau_2, \ldots, \tau_{k-1}$ can be found by maximizing the preceding objective function over the region $\tau_1 < \tau_2 < \cdots < \tau_{k-1}$.

5.5.1 Exponential distribution

Miller and Nelson [147] considered the optimal choice of the stress changing time for a simple step-stress life test, when the distribution of the lifetime at each stress level is an exponential distribution with mean lifetime θ_1 and θ_2. The authors assumed that the lifetime satisfies CEM assumptions under the step-stress pattern and used the log-linear link function, i.e.,

$$\ln \theta_i = \alpha + \beta x_i$$

to describe the relationship between the lifetime distribution and the corresponding stress level. Based on the complete sample, the log-likelihood function of α and β is given by

$$l(\alpha, \beta) = -n_1(\alpha + \beta x_1) - (n - n_1)(\alpha + \beta x_2) - T_1(\tau_1)e^{-(\alpha + \beta x_1)} - T_2(\tau_1)e^{-(\alpha + \beta x_2)},$$

where $T_1(\tau_1) = \sum_{i=1}^{n_1} t_{i:n} + (n - n_1)\tau_1$ and $T_2(\tau_1) = \sum_{i=n_1+1}^{n}(t_{i:n} - \tau_1)$. The Fisher information matrix of α and β can be computed taking the expectation of the negative of double and mixed partial derivatives of $l(\alpha, \beta)$ and it is given by

$$I(\alpha, \beta) = n \begin{pmatrix} A_1(\tau_1) + A_2(\tau_1) & x_1 A_1(\tau_1) + x_2 A_2(\tau_1) \\ x_1 A_1(\tau_1) + x_2 A_2(\tau_1) & x_1^2 A_1(\tau_1) + x_2^2 A_2(\tau_1) \end{pmatrix}, \tag{5.33}$$

where $A_1(\tau_1) = 1 - e^{-\tau_1/\theta_1}$ and $A_2(\tau_1) = e^{-\tau_1/\theta_1}$. Miller and Nelson [147] used the C-optimal criterion and obtained the optimal choice of τ_1 by minimizing

$$
\begin{aligned}
\phi_c(\tau_1) &= n \operatorname{AVar}\left(\ln \hat{\theta}_1\right) \\
&= n(1 \quad x_0)\,(I(\alpha, \beta))^{-1}\,(1 \quad x_0)' \\
&= \frac{(x_1 - x_0)^2 A_1(\tau_1) + (x_2 - x_0)^2 A_2(\tau_1)}{(x_1 - x_2)^2 A_1(\tau_1) A_2(\tau_1)} \\
&= \frac{(1 + \xi)^2}{A_1(\tau_1)} + \frac{\xi^2}{A_2(\tau_1)},
\end{aligned}
$$

where $\xi = \frac{x_1 - x_0}{x_2 - x_1}$ is the amount of extrapolation. The following theorem provides the C-optimal choice of the stress changing time for a simple SSLT. The proof of the theorem can be found in Miller and Nelson [147].

Theorem 5.5.1. *The C-optimal time is given by* $\tau_{1c} = \theta_1 \ln \frac{2\xi+1}{\xi}$.

Note that the optimal value of τ_1 depends on the unknown parameter θ_1. One needs some prior idea about the value of θ_1 to find the optimal choice of the stress changing time. Alternatively, a pilot survey may be conducted to estimate the unknown parameter θ_1 and use that estimate to find the optimal choice of τ_1.

Bai et al. [148] extended the results of Miller and Nelson [147] by incorporating the Type-I censoring scheme. Bai et al. [148] assumed that the experiment is terminated at the prefixed time η. Under the CEM assumption and log-liner link function, the Fisher information matrix is given by Eq. (5.33) with $A_1(\tau_1) = 1 - e^{-\tau_1/\theta_1}$ and $A_2(\tau_1) = e^{-\tau_1/\theta_1}\left(1 - e^{-(\eta-\tau_1)/\theta_2}\right)$. The following theorem provides the optimal choice of the stress changing time under the Type-I censoring. The interested readers are referred to Bai et al. [148] for the proof of the theorem.

Theorem 5.5.2. *The C-optimum stress changing time,* τ_{1c}, *is the unique solution of the following equation:*

$$
\left(\frac{A_1(\tau_1)}{A_2(\tau_1)}\right)^2 \times \frac{A_2(\tau_1) + \theta_1\left(1 - A_1(\tau_1) - A_2(\tau_1)\right)/\theta_2}{1 - A_1(\tau_1)} = \left(\frac{1+\xi}{\xi}\right)^2.
$$

Note that for the Type-I censoring scheme the optimal choice of the stress changing time depends on both the parameters θ_1 and θ_2. Prior information regarding the values of θ_1 and θ_2 is needed to obtain the value of τ_{1c}.

Now consider the simple step-stress life test, where the stress is changed as soon as the rth failure occurs. Here r is a prefixed integer and is less than n. Miller and Nelson [147] referred to this simple step-stress life test as the simple failure-step test. The author addressed the issue of the optimal choice of r under the CEM and exponential lifetimes. Note that to find the optimal value of r, the search needs to be confined on the set $\{1, 2, \ldots, n-1\}$, which makes it a discrete optimization problem. As before, assume that the mean lifetimes at the first and second stress levels are θ_1 and θ_2, respectively. Based on complete sample, the log-likelihood function of α and

β under the assumption of the log-linear link function can be expressed as follows:

$$l(\alpha, \beta) = -r(\alpha + \beta x_1) - (n - r)(\alpha + \beta x_2) - T_1 e^{-(\alpha + \beta x_1)} - T_2 e^{-(\alpha + \beta x_2)},$$

where $T_1 = \sum_{i=1}^{r} t_{i:n} + (n - r)t_{r:n}$ and $T_2 = \sum_{i=r+1}^{n} (t_{i:n} - t_{r:n})$. Hence the Fisher information matrix is given by

$$I(\alpha, \beta) = n \begin{pmatrix} A_1(r) + A_2(r) & x_1 A_1(r) + x_2 A_2(r) \\ x_1 A_1(r) + x_2 A_2(r) & x_1^2 A_1(r) + x_2^2 A_2(r) \end{pmatrix}, \tag{5.34}$$

where $A_1(r) = r$ and $A_2(r) = n - r$. Miller and Nelson [147] used the C-optimal criterion and obtained the optimal choice of r by minimizing

$$\phi_c(r) = n \, \text{AVar}\left(\ln \hat{\theta}_1\right) = \frac{(1 + \xi)^2}{A_1(r)} + \frac{\xi^2}{A_2(r)}$$

The following theorem provides the C-optimal choice of the stress changing time for a simple step-stress life test. The proof of the theorem can be found in Miller and Nelson [147].

Theorem 5.5.3. *The C-optimal value of r is given by $r_{1c} = \left[n \frac{1+\xi}{2\xi+1} \right]$.*

Note that the optimal value of r does not depend on the parameters θ_1 or θ_2. It only depends on the stress levels used and the normal stress level.

Bai et al. [148] and Kundu and Balakrishnan [141] extended the results of Miller and Nelson [147] for the Type-II censoring scheme. They assumed the experiment is terminated as soon as the sth failure occurs, where $s > r$. For a given s, the search of the optimal value of r needs to be done over the set $\{1, 2, \ldots, s-1\}$. The authors studied the optimal choice of r for given s and n. Under the CEM assumption and log-liner link function, the Fisher information matrix is given by Eq. (5.34) with $A_1(r) = r$ and $A_2(r) = s - r$. The following theorem provides the optimal choice of the stress changing time under Type-I censoring. The interested readers are referred to Bai et al. [148] for the proof of the theorem.

Theorem 5.5.4. *The C-optimum value of r is $r_c = \left[\frac{(n-s)(1+\xi)}{2\xi+1} \right]$.*

Kundu and Balakrishnan [141] considered A-optimal and D-optimal criteria. Their findings can be summarized in the following theorems.

Theorem 5.5.5. *A-optimal value of r can be found by minimizing $\phi_A(r) = \frac{\theta_1^2}{r} + \frac{\theta_2^2}{s-r}$.*

Theorem 5.5.6. *D-optimal value of r can be found by minimizing $\phi_D(r) = \frac{\theta_1^2 \theta_2^2}{r(n-r)}$, i.e., the D-optimal value of r is $\frac{s}{2}$ if s is an even integer; otherwise it will be either $\left[\frac{s}{2} \right]$ or $\left[\frac{s}{2} \right] + 1$, whichever minimizes $\frac{1}{r(n-r)}$.*

Note that the C-optimal and the D-optimal values of r do not depend on the parameters θ_1 and θ_2, but the A-optimal choice of r depends on the values of θ_1 and θ_2. However, if the ratio of θ_1 to θ_2 is known, i.e., $\theta_1 = c\theta_2$ for some known real number c, the A-optimal value of r can be found by minimizing $\phi_A = \theta_1^2 \left(\frac{c^2}{r} + \frac{1}{s-r} \right)$. In this case the A-optimal value of r is independent of parameters.

Khamis and Higgins [117] extend the results of Bai et al. [148] for the three-step step-stress tests. Similar to the log-linear link function, Khamis and Higgins [117] considered the log-quadratic link function. Optimum stress changing times for the multiple step-stress life test was considered by Khamis [149] using the log-linear link function and several stress variables. They used the C-optimal criterion to obtain the optimal stress changing time. The optimal stress changing time does not exist in explicit form for both the link functions and one needs to use a numerical technique to compute the same. As the steps of finding optimal stress changing time are same as before, we do not discuss it here. The interested readers are referred to the original articles by Khamis and Higgins [117] and Khamis [149].

Balakrishnan et al. [30] considered the simple SSLT, where the lifetimes are exponentially distributed with means θ_1 and θ_2 under the stress levels s_1 and s_2, respectively. It is assumed that the stress level is changed to s_2 from s_1 at the prefixed time τ_1 and the experiment is terminated as soon as the rth failure occurs. It is noted that the MLEs of θ_1 and θ_2 exist if there is at least one failure at each stress level. Kateri et al. [156] obtained the optimal change point τ_1 by minimizing the probability of nonexistence of MLEs of θ_1 and θ_2. Note that the MLE of θ_1 does not exist if the first failure occurs after time τ_1, and that of θ_2 does not exist if the rth failure occurs before time τ_1. Hence the probability of nonexistence of MLEs of θ_1 and θ_2 can be expressed as

$$P_{11} = (1 - F(\tau_1))^n \quad \text{and} \quad P_{12} = \sum_{j=r}^{n} \binom{n}{j} F^j(\tau_1)(1 - F(\tau_1))^{n-j},$$

respectively. Therefore, the probability of nonexistence of the MLE of θ_1 or θ_2 is

$$P_1 = P_{11} + P_{12} = (1 - F(\tau_1))^n + \sum_{j=r}^{n} \binom{n}{j} F^j(\tau_1)(1 - F(\tau_1))^{n-j},$$

as the two events are disjoint due to the exponentially distributed lifetimes. The following theorem states the optimal choice of stress changing time. The interested readers are referred to Kateri et al. [156] for more details.

Theorem 5.5.7. *In a simple SSLT, the probability of nonexistence of MLEs of θ_1 or θ_2 has a global minimum with respect to τ_1 at*

$$\tau_1^* = \theta_1 \ln \left(\left(\frac{n-1}{r-1} \right)^{-1/(r-1)} + 1 \right).$$

Kateri et al. [156] also studied two sample problems and showed that the probability of nonexistence of MLEs in the case of two samples never exceeds that for one equivalent sample. Unfortunately, for two samples the probability of nonexistence of MLEs of θ_1 or θ_2 does not possess a global finite two-dimensional minimum with respect to the stress changing times for both the samples.

Gouno et al. [157] and Han et al. [158] addressed the issue of the optimal design for a multiple step-stress life test with equispaced step times τ, where τ denotes the duration of each stress level. The authors assumed that the data are progressive Type-I censored. The problem can be described as follows. Let $c_1, c_2, \ldots, c_{k-1}$ be fixed integers. The test starts with $N_1 = n$ items at the stress level s_1. At the time τ, c_1 surviving items are removed from the test and the stress level is changed to s_2. The test continues with $N_2 = n - c_1 - n_1$ units. At the time 2τ, c_2 items are selected randomly from remaining surviving items and removed from the test. The stress level is changed to s_3 and the test continues with $N_3 = n - (n_1 + n_2) - (c_1 + c_2)$ items. This procedure is followed and finally at the time $k\tau$, all the surviving $c_k = n - \sum_{i=1}^{k} n_i - \sum_{i=1}^{k-1} c_i$ items are withdrawn from the test, thereby terminating the test. Here n_i is the number of failures during the stress level s_i. Under the assumption of an exponentially distributed lifetime under each stress level, CEM, and the log-linear link function, the Fisher information matrix is given by

$$I(\alpha, \beta) = n \begin{pmatrix} \sum_{i=1}^{k} A_i(\tau) & \sum_{i=1}^{k} A_i(\tau)x_i \\ \sum_{i=1}^{k} A_i(\tau)x_i & \sum_{i=1}^{k} A_i(\tau)x_i^2 \end{pmatrix}, \tag{5.35}$$

where $A_i(\tau) = \left(1 - \frac{1}{n}\sum_{j=1}^{i-1} c_j e^{\tau \sum_{l=1}^{j}(1/\theta_l)}\right) e^{-\tau \sum_{j=1}^{i-1}(1/\theta_j)} \left(1 - e^{-\tau/\theta_i}\right)$. Note that $A_i(\tau)$ can be negative for some values of τ, which gives rise to disconcerting anomalies, such as a negative variance function or a negative determinant of the Fisher information matrix. Clearly, this complexity is introduced by the progressive censoring scheme due to the nonexistence of the appropriate censoring scheme. For this reason, the authors suggested confining the search for optimal τ in the region $C_\tau = \{\tau : A_i(\tau) > 0, i = 2, 3, \ldots, k\}$, which ensures that the mean of N_i is greater than zero for all $i = 2, 3, \ldots, k$, i.e., on an average the number of surviving items at the end of any stage exceeds the number of items to be progressively censored at that time. The authors used C-optimality and D-optimality criteria and proved the existence of an optimal value of τ for the simple step-stress life test.

Balakrishnan and Han [159] pointed out that the expectation of N_i is greater than zero for $\tau \in C_\tau$, but N_i may not be greater than zero for each sample. To overcome this issue, Balakrishnan and Han [159] proposed to use the following censoring scheme. Let $\pi_1^*, \pi_2^*, \ldots, \pi_{k-1}^*$ be prefixed proportions, where $0 \le \pi_i^* < 1$ for $i = 1, 2, \ldots, k - 1$. The test starts with $N_1 = n$ items at the stress level s_1. At the time τ, $c_1 = (N_1 - n_1)\pi_1^*$ surviving items are removed from the test and the stress level is changed to s_2. The test continues with $N_2 = n - c_1 - n_1$ units. At the time 2τ, $c_2 = (N_2 - n_2)\pi_2^*$ items are selected randomly from the remaining surviving items and removed from the test. The stress level is changed to s_3 and the test continues with $N_3 = n - (n_1 + n_2) - (c_1 + c_2)$ items. This procedure is followed and finally at the time $k\tau$, all the surviving $c_k = n - \sum_{i=1}^{k} n_i - \sum_{i=1}^{k-1} c_i$ items are withdrawn from the test, thereby terminating the test. Under the assumption of exponentially distributed lifetime under each stress level, the CEM, and log-linear link function, the Fisher information matrix of α and β for this modified censored data is given by Eq. (5.35) with $A_i(\tau) = \left(1 - e^{-\tau/\theta_i}\right) e^{-\tau \sum_{j=1}^{i-1}(1/\theta_j)} \prod_{j=1}^{i-1}\left(1 - \pi_j^*\right)$, which is greater than zero for all values of τ. Consequently there is no restriction on the search region for the optimal τ for this modified censoring scheme. Balakrishnan and Han [159] considered A, C, and D-optimality criteria and proved the existence of the optimal stress duration for the simple step-stress life test. Note that even under this modified scheme the test can be terminated prematurely, i.e., before reaching the last stress level. Hence Balakrishnan and Han [159] also addressed the conditional inference conditioning on the event that $N_k > 0$. The conditional Fisher information matrix of α and β conditioning on $N_k > 0$ is given by Eq. (5.35), with

$$A_i(\tau) = \frac{1}{1 - H_1^n(\tau)}((1 - V_i\tau))F_i(\tau) + \tau(1 - H_i(\tau))V_i(\tau)e^{\alpha+\beta x_i} \prod_{j=1}^{i-1} S_j(\tau)(1 - \tau_j^*),$$

$$H_i(\tau) = \begin{cases} F_i(\tau) + (1 - F_i(\tau))H_{i+1}^{1-\pi_i^*}(\tau) & \text{if } i = 1, 2, \ldots, k-1 \\ 0 & \text{if } i = k, \end{cases}$$

$$V_i(\tau) = \begin{cases} \dfrac{H_1^{n-1}(\tau)}{\prod_{j=1}^{i-1} H_{j+1}^{\pi_j^*}(\tau)} & \text{if } i = 1, 2, \ldots, k-1 \\ 0 & \text{if } i = k, \end{cases}$$

$$F_i(\tau) = 1 - e^{-\tau/\theta_i}.$$

The optimal choice of the stress duration can be found with respect to appropriate criterion using numerical techniques. For the detailed proofs and computations, the interested readers are referred to Balakrishnan and Han [159].

Xie et al. [52] considered the optimal change point problem for a simple SSLT when the data is progressively Type-II censored. Under the exponential lifetimes and the CEM assumptions, the authors have suggested using $\mathrm{MSE}(\hat{\theta}_1) + \mathrm{Var}(\theta_2)$ as the objective function to obtain the optimal choice of stress changing time for a given progressive censoring scheme, where $\hat{\theta}_i$ is the MLE of θ_i, $i = 1, 2$. The motivations for this objective function are as follows. $\hat{\theta}_1$ is a biased estimator of θ_1, but $\hat{\theta}_2$ is an unbiased estimator of θ_2, and $\mathrm{Cov}(\hat{\theta}_1, \hat{\theta}_2) = 0$. Note that the same objective function can be used to compute the optimal progressive censoring scheme for a given stress changing time.

The optimal simple SSLT was considered by Ganguly and Kundu [133] when there are competing risks of failures. The authors assumed that the stress level is changed to s_2 from the initial stress level s_1 at the time when the rth failure occurs. The experiment is terminated at the time of the sth failure. Here r and s are prefixed integers satisfying $1 \leq r < s \leq n$. The authors, for simplicity, assumed that there are two causes of failure under each stress level, though the results can be extended for more than two causes. It is assumed that the lifetime of the items has an exponential distribution with means θ_{11} and θ_{12} under the stress levels s_1 for the first and second mode of failure, respectively. Similarly, the lifetime of an item is exponentially distributed with means θ_{21} and θ_{22} under the stress levels s_2 for the first and second mode of failure, respectively. It is also assumed that the lifetime under the step-stress pattern follows

CEM assumptions for each cause of failure. The authors modeled the competing risks using the latent failure time model of Cox [122]. The Fisher information matrix of θ_{11}, θ_{12}, θ_{21}, and θ_{22} under this set-up is given by

$$I(\theta_{11}, \theta_{12}, \theta_{21}, \theta_{22}) = \text{Diag}\left(\frac{rp_1}{\theta_{11}^2}, \frac{r(1-p_1)}{\theta_{12}^2}, \frac{(s-r)p_2}{\theta_{21}^2}, \frac{(s-r)(1-p_2)}{\theta_{22}^2}\right),$$

where $p_1 = \theta_{12}/(\theta_{11} + \theta_{12})$ and $p_2 = \theta_{21}/(\theta_{21} + \theta_{22})$. The objective functions of the D-optimality and the A-optimality are given by

$$\phi_D(r) = \frac{r(s-r)}{(\theta_{11} + \theta_{12})(\theta_{21} + \theta_{22})\sqrt{\theta_{11}\theta_{12}\theta_{21}\theta_{22}}}$$

and

$$\phi_A(r) = \frac{\theta_{11}\theta_{12}}{r} + \frac{\theta_{21}\theta_{22}}{s-r},$$

respectively. The D-optimal and A-optimal values of r can be found by minimizing $\phi_D(r)$ and $\phi_A(r)$, respectively, with respect to $r \in \{2, 3, \ldots, s-2\}$. The D-optimal value of r is $s/2$ if s is an even integer; otherwise it is given by $(r-1)/2$ or $(r+1)/2$, whichever produces a smaller value of $\phi_D(r)$. To find the A-optimal value of r, one needs to compute the values of $\phi_A(r)$ for all possible values of r, and the optimal value of r is the one that yields the smallest value of $\phi_A(r)$.

Bai and Chung [108] assumed the simple SSLT under the assumption of the tampered random variable model (TRVM), where the data is Type-I censored with the censoring time η. Under the assumption of the exponentially distributed lifetime at the first stress level with the mean $1/\lambda$, the authors obtained the optimal stress changing time with respect to the D-optimality criterion and also provided the asymptotic variance of the MLE of β. The former objective function is useful when one wants to estimate both the acceleration factor and hazard rate at the used condition. The latter objective function is useful when the aim of the experiment is to extrapolate the data from the accelerated condition to the used condition. Under the assumption of the TRVM, the CDF of the lifetime under the step-stress pattern is given by

$$f_{\tilde{T}}(t) = \begin{cases} \lambda e^{-\lambda x} & \text{if } t \leq \tau_1 \\ \beta\lambda e^{-\lambda(\tau_1 + \beta(t-\tau_1))} & \text{if } t > \tau_1 \\ 0 & \text{otherwise}, \end{cases}$$

where β is the acceleration factor. In this case the Fisher information matrix of β and λ is given by

$$I(\beta, \lambda) = n\begin{pmatrix} \frac{p_2}{\beta^2} & \frac{p_2}{\beta\lambda} \\ \frac{p_2}{\beta\lambda} & \frac{p_1+p_2}{\lambda^2} \end{pmatrix},$$

where $p_1 = 1 - e^{-\lambda \tau_1}$ and $p_2 = e^{-\lambda \tau_1} \left(1 - e^{-\beta \lambda (\eta - \tau_1)}\right)$. The objective function for the D-optimal criterion is given by

$$\phi_D(\tau_1) = \frac{\beta^2 \lambda^2}{n^2} \times \frac{1}{1 - q_0^{\tau_1/\eta}} \times \frac{1}{q_0^{\tau_1/\eta}(1 - q_1^{1-\tau_1/\eta})},$$

where $q_i = 1 - p_i$ for $i = 0, 1$, $p_0 = 1 - e^{-\lambda \eta}$, and $p_1 = 1 - e^{-\beta \lambda \eta}$. The asymptotic variance of MLE of β can be expressed as

$$\text{AVar}(\hat{\beta}) = \frac{\beta^2}{n}\left(\frac{1}{1 - q_0^{\tau_1/\eta}} + \frac{1}{q_0^{\tau_1/\eta}(1 - q_1^{1-\tau_1/\eta})}\right).$$

The following theorems provide the optimal values of τ_1 with respect to these objective functions. For the details of the proofs, the readers are referred to the original article by Bai and Chung [108].

Theorem 5.5.8. *The optimal change time with respect to the D-optimality criterion is the unique solution of*

$$q_0^{\tau_1/\eta}\left(1 - q_1^{1-\tau_1/\eta}\right) - \left(1 - q_0^{\tau_1/\eta}\right)\left(1 + \left(\frac{\ln q_1}{\ln q_0} - 1\right)q_1^{1-\tau_1/\eta}\right) = 0.$$

Theorem 5.5.9. *The optimal change time minimizing AVar$(\hat{\beta})$ is the unique solution of the equation*

$$\left(\ln q_0 - q_1^{1-\tau_1/\eta} \ln(q_0/q_1)\right)\left(1 - q_0^{\tau_1/\eta}\right)^2 - q_0^{2\tau_1/\eta}\left(1 - q_1^{1-\tau_1/\eta}\right)^2 \ln q_0 = 0.$$

5.5.2 Other distributions

In this section, we will briefly mention the literature that addresses the issues of the optimal SSLT under other distributions. In most of the cases the steps that may be used to find the optimal design are similar to that of the exponential distribution as discussed in the previous subsection. However, due to the complicated expressions, it is very difficult to obtain the theoretical results in these cases and most of the results are numerical in nature. Hence, instead of providing the details, here we will describe these problems, which have been addressed by several authors.

The optimal choice of stress changing points for the simple SSLT and for the multiple SSLT was addressed by Bai and Kim [160] and Chandra et al. [161], respectively, when the distribution of lifetime is Weibull with the same shape parameter and different scale parameters at the different stress levels. The authors assumed that the lifetime under the step-stress pattern follows the CEM assumptions. The authors have assumed that the scale parameter of the Weibull distribution at each stress level is a log-linear function of that stress level. Assuming that the data are Type-I censored, the authors minimized the asymptotic variance of the pth quantile of the lifetime at the normal stress level to find the optimal choice of the stress changing time.

Tang et al. [150] considered the optimality issue of the simple and multiple SSLT for the Weibull distribution under the assumption of TFRM. The authors minimized the asymptotic variance of the log of the pth quantile of the lifetime at the normal stress level to obtain the optimal stress changing time when the data is Type-I censored. Under the same set-up, Alhadeed and Yang [162] obtained the optimal stress changing time by minimizing the asymptotic variance of the pth quantile when the complete data is available.

The optimal stress changing time of a multiple-steps SSLT for independent competing risks was analyzed by Lui and Qui [44]. The authors used the latent failure time model by Cox [122], where the latent failure time has a Weibull distribution at each stress level. It is also assumed that the lifetime due to each cause of failure satisfies the CEM assumption under the whole step-stress pattern. The shape parameter is assumed to be constant across the stress levels for each cause of failure, where the scale parameter varies with the stress level for each mode of failure. A log-linear relationship is assumed between the scale parameters and the stress levels. The author used the D-optimality criterion and the negative of the asymptotic variance of the pth quantile of lifetime at the normal stress level and maximized them to obtain the optimum design of SSLT.

Bai et al. [115] considered optimality issues of a simple SSLT under the assumption of log-normal lifetime at the normal used condition. It is assumed that the lifetime satisfies the assumption of the TRVM and the data are Type-I censored.

> **Open Problem:** Kateri et al. [156] used the probability of nonexistence of the MLE as the objective function and minimized it for Type-II censored data to obtain the optimal design of the SSLT when the lifetimes follow the CEM assumptions under the exponential distribution at each stress level. This objective function can be used for other censoring schemes, other models and for other lifetime distributions.
>
> **Open Problem:** Note that the optimal choices of the stress changing times are the functions of the model parameters in most of the cases and the model parameters are usually unknown. Practitioners perform a pilot survey and use the estimate of the parameters obtained from the pilot survey to find the optimal design of the SSLT. Hence it is quite important to perform a sensitivity analysis to check the adequacy of the method with respect to small changes in the values of the parameters.

5.6 Further reading

We should mention that the amount of literature related to the step-stress model which has appeared in the last few decades is quite significant. We could not discuss in detail all the related materials in this monograph because of the limitations of space. Finally we will wrap up this chapter and the monograph by mentioning briefly several isolated works which have appeared in the literature since the inception of the step-stress model and indicating some open problems.

So far we have discussed different parametric models associated with the step-stress models and their inferential issues. Shaked and Singpurwalla [163] considered the multiple step-stress model using the nonparametric approach. They have not assumed any specific parametric model of the lifetime distribution of the experimental units. The basic idea of this nonparametric model is that of the shock models and of wear processes. The nonparametric model introduced by Shaked and Singpurwalla [163] generalizes the CEM and TFRM in some cases. The authors proposed a consistent estimator for the lifetime distribution under this set-up. Recently, Hu et al. [164] proposed a nonparametric proportional hazard model and provided an estimation of the lifetime distribution under normal conditions; see also Tyoskin and Krivolapov [165] in this respect. It may be mentioned that recently a proportional reversed hazard model has gained some popularity in the statistical literature because of its real-life applications; see for example Block et al. [139] for an introduction to the proportional reversed hazard model. It will be interesting to develop the necessary statistical inference for the proportional reversed hazard model in the step-stress set-up.

Dharmadhikari and Rahman [166] proposed a multiple step-stress model based on the assumption that the lifetime of the experimental units at a given stress level follows a Weibull or log-normal distribution and its scale parameter depends upon the present level as well as the age at the entry in the present level. They provided a detailed analysis for three step-stress models. Recently, Shemehsavar and Amini [167] provided a new multiple step-stress model based on a bivariate Wiener process; see also Pan and Balakrishnan [168] and Pan et al. [169], who had proposed some degradation models based on the Wiener and Gamma processes.

Huang [170] considered Cox's [22] proportional hazard model based on the multiple step-stress model. It is assumed that the hazard function at the stress level s satisfies the Cox's proportional hazard model assumption, i.e.,

$$\lambda(t; s) = \lambda_0(t) \exp(\beta s).$$

It is further assumed that the baseline hazard function $\lambda_0(t)$ has the following form

$$\lambda_0(t) = \gamma_0 + \gamma_1 t.$$

Based on this assumption the CDF at the stress level s can be obtained as

$$F(t; s) = \begin{cases} 0 & \text{if } t < 0 \\ 1 - \exp\left[-\left(\gamma_0 t + \frac{\gamma_1}{2} t^2\right) \exp(\beta s)\right] & \text{if } t \geq 0. \end{cases}$$

Now based on the CEM assumptions, Huang [170] provided the MLEs of the unknown parameters for the multiple step-stress model, and also provided an optimal design based on the A-optimality and D-optimality criteria. It will be interesting to develop the statistical inference for other baseline hazard functions as well.

Recently Liu et al. [171] considered the following problem. Suppose there are k stress levels $s_1 < s_2 < \cdots < s_k$ and $k - 1$ prefixed time points $\tau_1 < \tau_2 < \cdots < \tau_{k-1}$. Two groups of parallel systems of size n_1 and n_2 are tested simultaneously in the following manner. All the items of the first group are put initially at the stress level s_1, and this continues till the time τ_1. Then the stress is increased to s_2 and it continues till the time point τ_2 and so on. Finally at the time τ_{k-2}, the stress is increased to s_{k-1} and it continues till the experiment stops at τ_{k-1}. Similarly, all the items of the second group are put initially at the stress level s_2, and it continues till the time τ_1. The stress is then increased to s_2 which continues till τ_2 and so on. Finally at the time τ_{k-2}, the stress in increased to s_k and the experiment stops at τ_{k-1}. The authors provided nonparametric Bayesian estimators of the reliability functions corresponding to any components set based on the Markov chain Monte Carlo (MCMC) technique.

Some of the other work related to step-stress models can be found in Benavides [172], Frad and Li [173], Ginebra and Sen [174], Haghighi and Bae [175], Hong et al. [176], Li and Fard [177], Ma and Meeker [178], Xu and Fei [179], and the references cited there.

Bibliography

[1] W.B. Nelson, Accelerated Life Testing, Statistical Models, Test Plans and Data Analysis, John Wiley and Sons, New York, 1990.

[2] V.B. Bagdonavicius, M. Nikulin, Accelerated Life Models: Modeling and Statistical Analysis, Chapman and Hall CRC Press, Boca Raton, Florida, 2002.

[3] W.Q. Meeker, L.A. Escobar, Statistical Methods for Reliability Data, John Wiley and Sons, New York, 1998.

[4] J.F. Lawless, Statistical Models and Methods for Lifetime Data, second ed., John Wiley and Sons, New York, 2003.

[5] R.G. Miller, Survival Analysis, John Wiley and Sons, New York, 1988.

[6] L. J. Bain, M. Englehardt, Statistical Analysis of Reliability and Life-Testing Models: Theory and Methods, Marcel Dekker, New York, 1991.

[7] B. Epstein, Truncated life-test in exponential case, Ann. Math. Stat. 25 (1954) 555–564.

[8] A. Childs, B. Chandrasekar, N. Balakrishnan, D. Kundu, Exact likelihood inference based on Type-I and Type-II hybrid censored samples from the exponential distribution, Ann. Inst. Stat. Math. 55 (2003) 319–330.

[9] B. Chandrasekar, A. Childs, N. Balakrishnan, Exact inference for the exponential distribution under general Type-I and Type-II hybrid censoring, Nav. Res. Logist. 51 (2004) 994–1004.

[10] N. Balakrishnan, D. Kundu, Hybrid censoring: models, inferential results and applications, Comput. Stat. Data Anal. 57 (2013) 166–209.

[11] N. Balakrishnan, E. Cramer, The Art of Progressive Censoring: Applications to Reliability and Quality, Birkhäuser, Boston, 2014.

[12] N. Balakrishnan, Progressive censoring methodology: an appraisal (with discussion), Test 16 (2007) 211–296.

[13] D. Kundu, A. Joarder, Analysis of Type-II progressive hybrid censored data, Comput. Stat. Data Anal. 50 (2006) 2509–2528.

[14] A. Childs, B. Chandrasekar, N. Balakrishnan, Exact likelihood inference for the exponential parameter under progressive hybrid censoring, in: F. Vonta, M. Nikulin, N. Limnios, C. Huber-Carol (Eds.), Statistical Models and Methods for Biomedical and Technical System, Birkhäuser, Boston, 2008, pp. 319–330.

[15] N.M. Sediakin, On one physical principle in reliability theory, Tech. Cybern. 3 (1966) 80–87.

[16] W.B. Nelson, Accelerated life testing: step-stress models and data analysis, IEEE Trans. Reliab. 29 (1980) 103–108.

[17] P.K. Goel, Some estimation problems in the study of tampered random variables, Tech. Rep. Technical report no. 50, Department of Statistics, Carnegie-Mellon University, Pittsburgh, Pennsylvania, 1971.

[18] P.K. Goel, Consistency and asymptotic normality of maximum likelihood estimators, Scand. Actuar. J. 2 (1975) 109–118.

[19] M.H. DeGroot, P.K. Goel, Bayesian estimation and optimal design in partially accelerated life testing, Nav. Res. Logist. 26 (1979) 223–235.

[20] B.R. Rao, Equivalence of the tampered random variable and the tampered failure rate models in accelerated life testing for a class of life distribution having the 'setting the clock back to property', Commun. Stat. Theory Methods 21 (1992) 647–664.

[21] G.K. Bhattacharyya, Z. Soejoeti, A tampered failure rate model for step-stress accelerated life test, Commun. Stat. Theory Methods 18 (1989) 1627–1643.

[22] D.R. Cox, Regression models and life tables (with discussion), J. R. Stat. Soc. Ser. B 34 (1972) 187–202.

[23] M.T. Madi, Multiple step-stress accelerated life test: the tampered failure rate model, Commun. Stat. Theory Methods 22 (1993) 2631–2639.

[24] I.H. Khamis, J.J. Higgins, A new model for step-stress testing, IEEE Trans. Reliab. 47 (1998) 131–134.

[25] N. Sha, R. Pan, Bayesian analysis for step-stress accelerated life testing using Weibull proportional hazard model, Stat. Pap. 55 (2014) 715–726.

[26] H. Xu, Y. Tang, Commentary: The Khamis/Higgins model, IEEE Trans. Reliab. 52 (2003) 4–6.

[27] J.R. Drop, T.A. Mazzuchi, G.E. Fornell, L.R. Pollock, A Bayes approach to step-stress accelerated life testing, IEEE Trans. Reliab. 45 (1996) 491–498.

[28] N. Kannan, D. Kundu, N. Balakrishnan, Survival models for step-stress experiments with lagged effects, in: M. Nikulin, N. Limnios, N. Balakrishnan (Eds.), Advances in Degradation Modeling, Birkhauser, New York, 2010, pp. 355–369.

[29] C. Xiong, Inference on a simple step-stress model with Type-II censored exponential data, IEEE Trans. Reliab. 47 (1998) 142–146.

[30] N. Balakrishnan, D. Kundu, H.K.T. Ng, N. Kannan, Point and interval estimation for a simple step-stress model with Type-II censoring, J. Qual. Technol. 9 (2007) 35–47.

[31] S. Mitra, A. Ganguly, D. Samanta, D. Kundu, On simple step-stress model for two-parameter exponential distribution, Stat. Methodol. 15 (2013) 95–114.

[32] M. Kateri, N. Balakrishnan, Inference for a simple step-stress model with Type-II censoring and Weibull distributed lifetime, IEEE Trans. Reliab. 57 (2008) 616–626.

[33] L. Alkhalfan, Inference for a gamma step-stress model under censoring, Ph.D. thesis, McMaster University, Canada, 2012.

[34] A.A. Alhadeed, S. Yang, Optimal simple step-stress plan for cumulative exposure model using log-normal distribution, IEEE Trans. Reliab. 54 (2005) 64–68.

[35] T. Sun, Y. Shi, Estimation for Birnbaum-Saunders distribution in simple step-stress accelerated life test with Type-II censoring, Commun. Stat. Simul. Comput. 45 (2016) 880–901.

[36] M. Kamal, S. Zarrin, A.U. Islam, Step stress accelerated life testing plan for two parameter Pareto distribution, Reliab. Theory Appl. 8 (2013) 30–40.

[37] A. Arefi, M. Razmkhah, Optimal simple step-stress plan for Type-I censored data from geometric distribution, J. Iran. Stat. Soc. 12 (2013) 193–210.

[38] J. Klein, A. Basu, Accelerated life tests under competing Weibull causes of failure, Commun. Stat. Theory Methods 11 (1982) 2271–2286.

[39] J. Klein, A. Basu, Accelerated life testing under competing exponential failure distributions, IAPQR Trans. 7 (1982) 1–16.

[40] F. Pascual, Accelerated life test planning with independent Weibull competing risks with known shape parameter, IEEE Trans. Reliab. 56 (2007) 85–93.

[41] F. Pascual, Accelerated life test planning with independent Weibull competing risks, IEEE Trans. Reliab. 57 (2008) 435–444.

[42] N. Balakrishnan, D. Han, Exact inference for a simple step-stress model with competing risks for failure from exponential distribution under Type-II censoring, J. Stat. Plan. Inference 138 (2008) 4172–4186.

[43] D. Han, N. Balakrishnan, Inference for a simple step-stress model with competing risks for failure from the exponential distribution under time constraint, Comput. Stat. Data Anal. 54 (2010) 2066–2081.

[44] X. Liu, W.S. Qiu, Modeling and planning of step-stress accelerated life test with independent competing risks, IEEE Trans. Reliab. 60 (2011) 712–720.

[45] D. Han, D. Kundu, Inference for step-stress model with competing risks for failure from the generalized exponential distribution under Type-I censoring, IEEE Trans. Reliab. 64 (2015) 31–43.

[46] D. Han, Estimation in step-stress life tests with complementary risks from the exponentiated exponential distribution under time constraint and its applications to UAV data, Stat. Methodol. 38 (2015) 169–189.

[47] C. Xiong, G.A. Milliken, Step-stress life testing with random stress changing times for exponential data, IEEE Trans. Reliab. 48 (1999) 141–148.

[48] N. Balakrishnan, G. Iliopoulos, Stochastic monotonicity of the MLEs of parameters in exponential simple step-stress models under Type-I and Type-II censoring, Metrika 72 (2010) 89–109.

[49] N. Balakrishnan, Q. Xie, D. Kundu, Exact inference for a simple step-stress model from the exponential distribution under time constrain, Ann. Inst. Stat. Math. 61 (2009) 251–274.

[50] N. Balakrishnan, Q. Xie, Exact inference for a simple step-stress model with Type-I hybrid censored data from the exponential distribution, J. Stat. Plan. Inference 137 (2007) 3268–3290.

[51] N. Balakrishnan, Q. Xie, Exact inference for a simple step-stress model with Type-II hybrid censored data from the exponential distribution, J. Stat. Plan. Inference 137 (2007) 2543–2563.

[52] Q. Xie, N. Balakrishnan, D. Han, Exact inference and optimal censoring scheme for a simple step-stress model under progressive Type-II censoring, in: A.C. Arnold, N. Balakrishnan, J.M. Sarabai, R. Minguez (Eds.), Advances in Mathematical and Statistical Modeling, Birkhäuser, Boston, 2008, pp. 107–137.

[53] A. Ganguly, D. Kundu, S. Mitra, Bayesian analysis of simple step-stress model under Weibull lifetimes, IEEE Trans. Reliab. 64 (2015) 473–485.

[54] N. Balakrishnan, A synthesis of exact inferential results for exponential step-stress models and associated optimal accelerated life-tests, Metrika 69 (2009) 351–396.

[55] N. Balakrishnan, L. Zhang, Q. Xie, Inference for a simple step-stress model with Type-I censoring and lognormally distributed lifetimes, Commun. Stat. Theory Methods 38 (2009) 1690–1709.

[56] C.T. Lin, C.C. Chou, Statistical inference for a lognormal step-stress model with Type-I censoring, IEEE Trans. Reliab. 61 (2012) 361–377.

[57] A. Abdel-Hamid, E. K. Al-Hussaini, Estimation in step-stress accelerated life tests for the exponentiated exponential distribution with Type-I censoring, Comput. Stat. Data Anal. 53 (2009) 1328–1338.

[58] R. Wang, X. Xu, R. Pan, N. Sha, On parameter interference for a step-stress accelerated life test with geometric distribution, Commun. Stat. Theory Methods 41 (2012) 1796–1812.

[59] N.L. Johnson, S. Kotz, N. Balakrishnan, Continuous Univariate Distribution, vol. 1, second ed., John Wiley and Sons, New York, USA, 1994.

[60] D.J. Bartholomew, The sampling distribution of an estimate arising in life-testing, Technometrics 5 (1963) 361–372.

[61] G. Casella, R.L. Berger, Statistical Inference, Duxbury Press, Beltmont, CA, USA, 1990.

[62] S.M. Chen, G.K. Bhattacharyya, Exact confidence bound for an exponential parameter under hybrid censoring, Commun. Stat. Theory Methods 17 (1988) 1857–1870.

[63] D. Kundu, S. Basu, Analysis of competing risk models in presence of incomplete data, J. Stat. Plan. Inference 87 (2000) 221–239.

[64] N. Balakrishnan, G. Iliopoulos, Stochastic monotonicity of the MLE of exponential mean under different censoring schemes, Ann. Inst. Stat. Math. 61 (2009) 753–772.

[65] B. Efron, R. Tibshirani, An Introduction to the Bootstrap, Chapman and Hall CRC Press, Boca Raton, Florida, 1993.

[66] N. Balakrishnan, A. Rasouli, A.S. Farsipour, Exact likelihood inference based on an unified hybrid censoring sample from the exponential distribution, J. Stat. Comput. Simul. 78 (2008) 475–488.

[67] D.V. Lindley, Approximate Bayes method, Trab. Estad. 31 (1980) 223–237.

[68] D. Goldberg, Genetic Algorithm in Search, Optimization and Machine Learning, Addision-Wesley Professional, Reading, MA, 1989.

[69] S. Kirkpatrick, C.D. Gelatt Jr., M.P. Vecchi, Optimization by simulated annealing, Science 220 (1983) 671–680.

[70] R.D. Gupta, D. Kundu, Generalized exponential distribution, Aust. N. Z. J. Stat. 41 (1999) 173–188.

[71] G.S. Mudholkar, D.K. Srivastava, Exponentiated Weibull family for analyzing bathtub failure data, IEEE Trans. Reliab. 42 (1993) 299–302.

[72] G.S. Mudholkar, D.K. Srivastava, M. Freimer, The exponentiated Weibull family; a reanalysis of the bus motor failure data, Technometrics 37 (1995) 436–445.

[73] R.D. Gupta, D. Kundu, Closeness of gamma and generalized exponential distribution, Commun. Stat. Theory Methods 32 (2003) 705–721.

[74] R.D. Gupta, D. Kundu, Discriminating between Weibull and generalized exponential distribution, Comput. Stat. Data Anal. 43 (2003) 179–196.

[75] R.D. Gupta, D. Kundu, Generalized exponential distribution: existing methods and recent developments, J. Stat. Plan. Inference 137 (2007) 3537–3547.

[76] S. Nadarajah, The exponentiated exponential distribution: a survey, Adv. Stat. Anal. 95 (2011) 219–251.

[77] E.K. Al-Hussaini, M. Ahsanullah, Exponentiated Distributions, Atlantis Press, Paris, France, 2015.

[78] D. Samanta, D. Kundu, Comparison between order restricted and without order restricted inference of a simple step stress model, 2016, Technical Report, IIT Kanpur.

[79] A.K. Gupta, S. Nadarajah, Handbook of Beta Distribution and Its Applications, Marcel Dekker, New York, NY, 2004.

[80] P. Congdon, Applied Bayesian Modelling, second ed., Wiley, New York, NY, 2014.

[81] D. Kundu, R.D. Gupta, Closeness of gamma and generalized exponential distributions, Commun. Stat. Theory Methods 32 (2003) 705–721.

[82] D.H. Fearn, E. Nebenzahl, On the maximum likelihood ratio method of deciding between the Weibull and gamma distributions, Commun. Stat. Theory Methods 20 (1991) 577–593.

[83] K.O. Bowman, L.R. Shenton, Properties of Estimators for the Gamma Distribution, CRC Press, New York, USA, 1988,

[84] H.M. Srivastava, J. Choi, Series Associated with the Zeta and Related Functions, Kluwer Academic Publisher, Netherlands, 2001.

[85] D. Kundu, A. Manglick, Discriminating between the Weibull and log-normal distributions, Nav. Res. Logist. 51 (2004) 893–905.

[86] D. Kundu, A. Manglick, Discriminating between the gamma and log-normal distributions, J. Appl. Stat. Sci. 14 (2005) 175–187.

[87] D. Kundu, R.D. Gupta, A. Manglick, Discriminating between the log-normal and generalized exponential distributions, J. Stat. Plan. Inference 127 (2005) 213–227.

[88] A.A. Alhadeed, Models for step-stress accelerated life testing, Ph.D. thesis, Kansas State University, USA, 1998.

[89] A.K. Dey, D. Kundu, Discriminating between the log-normal and log-logistic distributions, Commun. Stat. Theory Methods 37 (2010) 280–292.

[90] P.W. Srivastava, R. Shukla, A log-logistic step-stress model, IEEE Trans. Reliab., 57 (2008), 431–434.

[91] A. Al-Masri, M.A. Ebrahem, Optimum time log-logistic cumulative exposure model using log-logistic distribution with known scale parameter, Aust. J. Stat. 38 (2009) 59–66.

[92] M.A. Ebrahem, A. Al-Masri, Optimum simple step-stress plan for log-logistic cumulative exposure model, Metron LXV (2007) 23–34.

[93] V. Pareto, Cours d'Economie Politique 2, F. Rouge, Lausanne, Switzerland, 1897.

[94] B.C. Arnold, Pareto Distribution, second ed., Chapman and Hall CRC Press, Boca Raton, Florida, 2015.

[95] Z.W. Birnbaum, S.C. Saunders, Estimation for a family of life distribution with applications to fatigue, J. Appl. Probab. 6 (1969) 328–347.

[96] Z.W. Birnbaum, S.C. Saunders, A new family of life distribution, J. Appl. Probab. 6 (1969) 319–327.

[97] A.F. Desmond, Stochastic models of failure in random environments, Can. J. Stat. 13 (1985) 1–28.

[98] V. Leiva, The Birnbaum-Saunders Distribution, first ed., Academic Press, Oxford, UK, 2016.

[99] R. Neal, Slice sampling, Ann. Stat. 31 (2003) 705–767.

[100] A. Xu, Y. Tang, Bayesian analysis of Birnbaum-Saunders distribution with partial information, Comput. Stat. Data Anal. 55 (2011) 2324–2333.

[101] M. Wang, X. Sun, C. Park, Bayesian analysis of Birnbaum-Saunders distribution via the generalized ratio of-uniforms method, Comput. Stat. 31 (2016) 207–225.

[102] N.L. Johnson, A.W. Kemp, S. Kotz, Univariate Discrete Distribution, third ed., John Wiley and Sons, New York, USA, 2005.

[103] A. Arefi, M. Razmkhah, Point and interval estimation for a simple step-stress model with Type-I censored data from the geometric distribution, 2011, unpublished manuscript.

[104] A. Arefi, M. Razmkhah, M. Borzadaran, Bayes estimation for a simple step-stress model with Type-I censored data from the geometric distribution, J. Stat. Res. Iran 8 (2011) 149–169.

[105] G. Gan, L.J. Bain, Distribution of order statistics for discrete parents with applications to censored sampling, J. Stat. Plan. Inference 44 (1995) 37–46.

[106] S. Kotz, N. Balakrishnan, N.L. Johnson, Continuous Multivariate Distribution, vol. 1, second ed., John Wiley and Sons, New York, USA, 2000.

[107] E. Gouno, N. Balakrishnan, Step-stress accelerated life test, in: N. Balakrishnan, C.R. Rao (Eds.), Handbook of Statistics 20: Advances in Reliability, Elsevier BV, Amsterdam, 2001, pp. 623–639.

[108] D.S. Bai, S.W. Chung, Optimal design of partially accelerated life test for the exponential distribution under Type-I censroing, IEEE Trans. Reliab. 41 (1992) 400–406.

[109] A.A. Abdel-Ghaly, A.F. Attia, M.M. Abdel-Ghani, The maximum likelihood estimates in step partially accelerated life tests for the Weibull parameters in censored data, Commun. Stat. Theory Methods 31 (2012) 551–573.

[110] A. Miele, S. Naqvi, A.V. Levy, Modified quasilinearization method for solving nonlinear equations, Tech. Rep. Technical report no. 78, Rice University, Houston, Taxes, 1970.

[111] A.A. Ismail, Estimation under failure-censored step-stress life test for the generalized exponential distribution parameters, Indian J. Pure Appl. Math. 4 (2014) 1003–1015.

[112] M.M.M. El-Din, S.E. Abu-Youssef, N.S.A. Ali, A.M.A. El-Raheem, Estimation in step-stress accelerated life tests for power generalized Weibull distribution with progressive censoring, Adv. Stat. 2015 (2015) doi:10.1155/2015/319051.

[113] Y. Shi, X. Shi, Estimation and optimal plan in step-stress partially accelerated life test model with progressive hybrid censored data from Pareto distribution, J. Phys. Sci. 20 (2015) 53–62.

[114] R.M. El-Sagheer, M. Ahsanullah, Statistical inference for a step-stress partially accelerated life test model based on progressively Type-II censored data from Lomax distribution, J. Appl. Stat. Sci. (2015) 307–323.

[115] D.S. Bai, S.W. Chung, Y.R. Chun, Optimal design of partially accelerated life tests for the lognormal distribution under Type-I censoring, Reliab. Eng. Syst. Safe. 40 (1993) 85–92.

[116] R. Wang, H. Fei, Uniqueness of the maximum likelihood estimate of the Weibull distribution tampered failure rate model, Commun. Stat. Theory Methods 32 (2003) 2321–2338.

[117] I.H. Khamis, J.J. Higgins, Optimum 3-step step-stress tests, IEEE Trans. Reliab. 45 (1996) 341–345.

[118] J.R. Drop, T.A. Mazzuchi, A general Bayes exponential inference model for accelerated life testing, J. Stat. Plan. Inference 119 (2004) 55–74.

[119] J. Beltrami, Exponential competing risk step-stress model with lagged effect, Int. J. Math. Stat. 16 (2015) 1–25.

[120] J. Beltrami, Weibull lagged effect step-stress model with competing risks, Commun. Stat. Theory Methods (2016) doi:10.1080/03610926.2015.1102283.

[121] A. Tsiatis, A nonidentifiability aspect of the problem of competing risks, Proc. Natl. Acad. Sci. U.S.A. 77 (1975) 20–22.

[122] D.R. Cox, The analysis of exponentially distributed lifetimes with two types of failures, J. R. Stat. Soc. Ser. B 21 (1959) 411–421.

[123] R.L. Prentice, J.D. Kalbfleish, A.V. Peterson Jr., N. Flurnoy, V.T. Farewell, N.E. Breslow, The analysis of failure time points in presence of competing risks, Biometrics 34 (1978) 541–544.

[124] D. Kundu, Parameter estimation of the partially complete time and type of failure data, Biom. J. 46 (2004) 165–179.

[125] M.J. Crowder, Classical Competing Risks, Chapman and Hall CRC Press, Boca Raton, Florida, 2001.

[126] M.J. Crowder, Multivariate Survival Analysis and Competing Risk, Chapman and Hall CRC Press, Boca Raton, Florida, 2012.

[127] J. Klein, A. Basu, Weibull accelerated life tests when there are competing causes os failure, Commun. Stat. Theory Methods, 10 (1981) 2073–2100.

[128] D.S. Bai, Y.R. Chun, Optimum simple step-stress accelerated life tests with competing causes of failure, IEEE Trans. Reliab. 40 (1991) 622–627.

[129] C.M. Kim, D.S. Bai, Analyses of accelerated life test data under two failure modes, Int. J. Reliab. Qual. Safe. Eng. 9 (2002) 111–125.

[130] P.W. Srivastava, R. Shukla, K. Sen, Optimum simple step-stress test with competing risks for failure using Khamis-Higgins model under Type-I censoring, Int. J. Oper. Res. Nepal 3 (2014) 75–88.

[131] F. Liu, Y. Shi, Inference for a simple step-stress model with progressively censored competing risks data from Weibull distribution, Commun. Stat. Theory Methods (2016) doi:10.1080/03610926.2016.1147585.

[132] C. Zhang, Y. Shi, M. Wu, Statistical inference for competing risks model in step-stress partially accelerated life tests with progressively Type-I hybrid censored Weibull life data, J. Comput. Appl. Math. 297 (2016) 65–74.

[133] A. Ganguly, D. Kundu, Analysis of simple step-stress model in presence of competing risks, J. Stat. Comput. Simul. 86 (2016) 1989–2006.

[134] A.P. Basu, J.K. Ghosh, Identifiability of distributions under competing risks and complementary risks model, Commun. Stat. Theory Methods 14 (1980) 1515–1525.

[135] J. Beltrami, Competing risks in the step-stress model with lagged effects, Ph.D. thesis, The University of Texas as San Antonio, USA, 2011.

[136] M. Wu, Y. Shi, Y. Sun, Inference for accelerated competing failure models from Weibull distribution under Type-I progressive hybrid censoring, J. Comput. Appl. Math. 263 (2014) 423–431.

[137] A.P. Basu, J. Klein, Some recent results in competing risks theory, in: Lecture Notes-Monograph, Series: Survival Analysis, vol. 2, 1982, pp. 216–229.

[138] A.P. Basu, Identifiability problems in the theory of competing and complementary risks—a survey, in: Statistical Distributions in Scientific Work, Reidel Publishing Company, Holland, 1981, pp. 335–348.

[139] H.W. Block, T.H. Savits, H.P. Singh, On the reversed hazard rate function, Probab. Eng. Inform. Sci. 12 (1998) 69–90.

[140] D. Kundu, R.D. Gupta, Characterizations of the proportional (reversed) hazard class, Commun. Stat. Theory Methods 38 (2004) 3095–3102.

[141] D. Kundu, N. Balakrishnan, Point and interval estimation for a simple step-stress model with random stress-change time, J. Probab. Stat. Sci. 7 (2009) 113–126.

[142] D. Samanta, D. Kundu, Analysis of Weibull step-stress model in presence of competing risks, 2016, Technical Report, IIT Kanpur.

[143] N. Balakrishnan, E. Beutner, M. Kateri, Order restricted inference for exponential step-stress models, IEEE Trans. Reliab. 58 (2009) 132–142.

[144] D. Samanta, A. Ganguly, D. Kundu, S. Mitra, Order restricted Bayesian inference for exponential simple step-stress model, Commun. Stat. Simul. Comput. doi:10.1080/03610918.2014.992540.

[145] M. Kateri, U. Kamps, N. Balakrishnan, A meta-analysis approach for step-stress experiments, J. Stat. Plan. Inference 139 (2009) 2907–2919.

[146] M. Kateri, U. Kamps, N. Balakrishnan, Multi-sample simple step-stress experiment under time constraints, Statistica Neerlandica 64 (2010) 77–96.

[147] R. Miller, W.B. Nelson, Optimum simple step-stress plans for accelerated life testing, IEEE Trans. Reliab. 32 (1983) 59–65.

[148] D.S. Bai, M.S. Kim, S.H. Lee, Optimum simple step-stress accelerated life test with censoring, IEEE Trans. Reliab. 38 (1989) 528–532.

[149] I.H. Khamis, Optimum M-step, step-stress design with K stress variables, Commun. Stat. Simul. Comput. 26 (1997) 1301–1313.

[150] Y. Tang, Q. Guan, P. Xu, H. Xu, Optimum design for Type-I step-stress accelerated life tests for two-parameter Weibull distribution, Commun. Stat. Theory Methods 41 (2012) 3863–3877.

[151] C. Zhang, Y. Shi, Estimation of extended Weibull parameters and acceleration factor in the step-stress accelerated life tests under an adaptive progressive hybrid censoring data, J. Stat. Comput. Simul. 86 (2016) 3303–3314.

[152] N. Balakrishnan, U. Kamps, M. Kateri, A sequential order statistics approach to step-stress testing, Ann. Inst. Stat. Math. 64 (2012) 303–318.

[153] U. Kamps, A concept of generalized order statistics, J. Stat. Plan. Inference 48 (1995) 1–23.

[154] D. Samanta, A. Ganguly, A. Gupta, D. Kundu, On multiple exponential step stress model under order restriction, 2017, Technical Report, IIT Kanpur, pp. 1–29.

[155] M. Kateri, U. Kamps, N. Balakrishnan, A meta-analysis approach for step-stress experiments, J. Stat. Plan. Inference 139 (2009) 2907–2919.

[156] M. Kateri, U. Kamps, N. Balakrishnan, Optimal allocation of change points in simple step-stress experiments under Type-II censoring, Comput. Stat. Data Anal. 55 (2011) 236–247.

[157] E. Gouno, A. Sen, N. Balakrishnan, Optimal step-stress test under progressive Type-I censoring, IEEE Trans. Reliab. 53 (2004) 388–393.

[158] D. Han, N. Balakrishnan, A. Sen, E. Gouno, Correction on "optimal step-stress test under progressive Type-I censoring", IEEE Trans. Reliab. 23 (2006) 103–122.

[159] N. Balakrishnan, D. Han, Optimal step-stress testing for progressive Type-I censored data for exponential distribution, J. Stat. Plan. Inference 139 (2009) 1782–1798.

[160] D.S. Bai, M.S. Kim, Optimum simple step-stress accelerated life tests for Weibull distribution and Type-I censoring, Nav. Res. Logist. 40 (1993) 193–210.

[161] N. Chandra, M.A. Khan, G. Gopal, Optimum quadratic step-stress accelerated life test plan for Weibull distribution under Type-I censoring, Int. J. Syst. Assur. Eng. Manag. (2016) 1–7, doi:10.1007/s13198-016-0473-8.

[162] A.A. Alhadeed, S. Yang, Optimal simple step-stress plan for Khamis-Higgins model, IEEE Trans. Reliab. 51 (2002) 212–215.

[163] M. Shaked, N.D. Singpurwalla, Inference for step-stress accelerated life tests, J. Stat. Plan. Inference 7 (1983) 295–306.

[164] C-H. Hu, R.D. Plante, J. Teng, Step-stress accelerated life tests: a proportional hazards-based non-parametric model, IIE Trans. 44 (2012) 754–764.

[165] O.I. Tyoskin, S.Y. Krivolapov, Nonparametric model for step-stress accelerated life testing, IEEE Trans. Reliab. 45 (1996) 346–350.

[166] A. Dharmadhikari, Md. M. Rahman, A model for step-stress accelerated life testing, Nav. Res. Logist. 50 (2003) 171–183.

[167] S. Shemehsavar, M. Amini, Failure inference and optimization for step stress model based on bivariate Wiener model, Commun. Stat. Simul. Comput. 45 (2016) 130–151.

[168] Z. Pan, N. Balakrishnan, Multiple steps step-stress accelerated degradation modeling based on Wiener and gamma process, Commun. Stat. Simul. Comput. 39 (2010) 1384–1402.

[169] Z. Pan, N. Balakrishnan, Q. Sun, Bivariate constant-stress accelerated degradation model and inference, Commun. Stat. Simul. Comput. 40 (2011) 247–257.

[170] W. Huang, Optimal and robust design of step-stress accelerated life testing experiments for proportional hazard models, Master's thesis, Faculty of Mathematics and Statistics, Brock University, St. Catharines, Ontario, 2015.

[171] B. Liu, Y. Shi, F. Zhang, X. Bai, Reliability nonparametric Bayesian estimation for the masked data of parallel systems in step-stress accelerated life tests, J. Comput. Appl. Math. (2017) 375–386.

[172] E.M. Benavides, Reliability model for step-stress and variable-stress situations, IEEE Trans. Reliab. 60 (2011) 219–233.

[173] N. Fard, C. Li, Optimum simple step-stress accelerated life test design for reliability prediction, J. Stat. Plan. Inference 139 (2009) 1799–1808.

[174] J. Ginebra, A. Sen, Minimax approach to accelerated life tests, IEEE Trans. Reliab. 47 (1998) 261–267.

[175] F. Haghighi, S.J. Bae, Reliability estimation from linear degradation and failure time data with competing risks under a step-stress accelerated degradation test, IEEE Trans. Reliab. 64 (2015) 960–971.

[176] Y. Hong, H. Ma, W.Q. Meeker, A tool for evaluating time-varying-stress accelerated life test plans with log-location-scale distributions step-stress accelerated life tests: a proportional hazards-based non-parametric model, IEEE Trans. Reliab. 59 (2010) 620–627.

[177] C. Li, N. Fard, Optimum bivariate step-stress accelerated life test for censored data, IEEE Trans. Reliab. 56 (2007) 77–84.

[178] H. Ma, W.Q. Meeker, Strategy for planning accelerated life tests with small sample sizes, IEEE Trans. Reliab. (2010) 610–619.

[179] H. Xu, H. Fei, Planning step-stress accelerated life tests with two experimental variables, IEEE Trans. Reliab. 56 (2007) 569–579.

Author Index

A

Abdel-Ghaly, A. A., 80, 84, 85, 87, 88
Abdel-Ghani, M. M., 80, 84, 85, 87, 88
Abdel-Hamida, A., 18, 47
Abu-Youssef, S. E., 90–92
Ahsanullah, M., 47, 93, 94
Al-Hussaini, E. K., 18, 47
Al-Masri, A., 62
Alhadeed, A. A., 15, 18, 59, 152
Ali, N. S. A., 90–92
Alkhalfan, L., 15, 18, 55, 57, 58
Amini, M., 153
Arefi, A., 15, 18, 69, 71
Arnold, B. C., 63
Attia, A. F., 80, 84, 85, 87, 88

B

Bae, S. J., 154
Bagdanavicius, V. B., 90
Bai, D. S., 80, 81, 83, 94, 106, 130,
    145–147, 150, 151, 152
Bai, X., 153
Bain, L. J., 4, 70
Balakrishnan, N., 4–7, 14, 15, 17–21, 23,
    25–28, 38, 41, 43–46, 59, 61, 62, 71, 74,
    98–101, 106, 113, 114, 116, 120, 121,
    129–133, 136–138, 142, 146–149,
    152, 153
Bartholmew, D. J., 21
Basu, A., 15, 106
Basu, A. P., 106, 125
Basu, S., 23
Beltarami, J., 100, 120, 122
Benavides, E. M., 154
Berger, R. L., 23
Beutner, E., 129, 137, 138
Bhattacharyya, G. K., 13, 15, 23, 95, 112
Birnbaum, Z. W., 65
Block, H. W., 125, 153
Borzadaran, M., 69, 71
Bowman, K. O., 55
Breslow, V. T. Farewell N. E., 105

C

Casella, G., 23
Chandra, N., 151
Chandrasekar, B., 4, 5, 7, 23
Chen, S. M., 23
Childs, A., 4, 5, 7, 23
Choi, J., 57
Chou, C. C., 18, 61
Chun, Y. R., 94, 106, 130, 152
Chung, S. W., 80, 81, 83, 94, 130,
    150–152
Congdon, P., 52
Cox, D. R., 13, 105, 107, 125, 133, 150,
    152, 153
Cramer, E., 6
Crowder, M. J., 105

D

DeGroot, M. H., 11, 15, 80–83
Desmond, A. F., 65
Dey, A. K., 62
Dharmadhikari, A., 153
Drop, J. R., 14, 99–101

E

Ebrahem, Al-Haj, 62
Ebrahem, M. A., 62
Efron, B., 25
El-Din, M. M. M., 90–92
El-Raheem, A. M. A., 90–92
El-Sagheer, R. M., 93, 94
Englehardt, M., 4
Epstein, B., 4
Escobar, L. A., 3

F

Fard, N., 154
Farsipour, A. S., 28
Fearn, D. H., 55
Fei, H., 97, 135, 154
Flurnoy, N., 105

Fornell, G. E., 14, 99–101
Freimer, M., 46

G

Gan, G., 70
Ganguly, A., 15, 17, 18, 26, 29–31, 35, 36,
    46, 106, 129, 133, 138–142, 149
Gelatt, C. D. Jr., 41
Ghosh, J. K., 106, 125
Ginebra, J., 154
Goel, P. K., 11, 15, 80–83
Goldberg, D., 41
Gopal, G., 151
Gouno, E., 74, 148
Guan, Q., 130, 152
Gupta, A., 138–142
Gupta, A. K., 49
Gupta, R. D., 46, 47
Gupta, R. D., 54, 59, 125

H

Haghighi, F., 154
Han, D., 15, 17, 28, 106, 114, 116, 120,
    121, 127, 148, 149
Higgins, J. J., 14, 97, 98, 130, 147
Hong, Y., 154
Hu, C-H., 153
Huang, W., 153

I

Iliopoulos, G., 17, 23
Islam, A. U., 15, 64, 65
Ismail, A. A., 89

J

Joarder, A., 7
Johnson, N. L., 19, 38, 59, 62, 69, 71

K

Kalbfleish, J. D., 105
Kamal, M., 15, 64, 65
Kamps, U., 129, 130, 136, 142, 147,
    152
Kannan, N., 14, 15, 17, 20, 21, 23, 25–27,
    99–101, 113, 147
Kateri, M., 15, 18, 38, 41, 43–46, 129,
    130, 136, 137, 138, 142, 147, 152

Kemp, A. W., 69
Khamis, I. H., 14, 97, 98, 130, 147
Khan, M. A., 151
Kim, C. M., 106
Kim, M. S., 130, 145–147, 151
Kirkpatrick, S., 41
Klein, J., 15, 106, 125
Kotz, S., 19, 38, 59, 62, 69, 71
Krivolopov, S. Y., 153
Kundu, D., 4, 6, 7, 14, 15, 17, 18, 20,
    21, 23, 25–27, 29–31, 35, 36, 46,
    47, 49, 52, 54, 59, 62, 99–101, 105, 106,
    113, 125, 129–133, 138–142, 146, 147,
    149

L

Lawless, J. F., 4
Lee, S. H., 130, 145–147
Leiva, V., 66
Levy, A. V., 85
Li, C., 154
Lin, C. T., 18, 61
Lindley, D. V., 34, 50, 68
Liu, B., 153
Liu, F., 106, 124
Liu, X., 15, 152

M

Ma, H., 154
Madi, M. T., 13, 15, 97
Manglick, A., 59
Mazzuchi, T. A., 14, 99–101
Meeker, W. Q., 3, 154
Miele, A., 85
Miller, R., 130, 144–146
Miller, R. G., 4
Milliken, G. A., 17, 129, 130
Mitra, S., 15, 17, 18, 26, 29–31, 35, 36,
    46, 129
Mudholkar, G. S., 46, 54

N

Nadarajah, S., 47, 49
Naqvi, S., 85
Neal, R., 68
Nebenzahl, E., 55
Nelson, W. B., 3, 10, 17, 130, 144–146

Ng, H. K. T., 15, 17, 20, 21, 23,
  25–27, 147
Nikulin, M., 90

P

Pan, R., 14, 18, 69, 98
Pan, Z., 153
Pareto, V., 63
Park, C., 68
Pascual, F., 15, 106
Peterson, Jr. A. V., 105
Plante, R. D., 153
Pollock, L. R., 14, 99–101
Prentice, R. L., 105

Q

Qiu, W. S., 15, 152

R

Rahman, Md. M., 153
Rao, B. R., 13
Rasouli, A., 28
Razmkhah, M., 15, 18, 69, 71

S

Samanta, D., 15, 18, 26, 29–31, 35, 36, 46,
  47, 49, 52, 54, 129, 138–142
Saunders, S. C., 65
Savits, T. H., 125, 153
Sen, A., 148, 154
Sen, K., 106
Sediakin, N. M., 10, 17
Sha, N., 14, 18, 69, 98
Shaked, M., 153
Shemehsavar, S., 153
Shenton, L. R., 55
Shi, X., 92, 93
Shi, Y., 15, 66–68, 92, 93, 106, 124,
  136, 153
Shukla, R., 62, 106
Singh, H. P., 125, 153

Singpurwala, N. D., 153
Soejoeti, Z., 13, 15, 95, 112
Srivastava, D. K., 46, 54
Srivastava, H. M., 57
Srivastava, P. W., 62, 106
Sun, Q., 153
Sun, T., 15, 66–68
Sun, X., 68

T

Tang, Y., 14, 68, 130, 152
Teng, J., 153
Tibshirani, R., 25
Tsiatis, A., 105
Tyoskin, O. I., 153

V

Vecchi, M. P., 41

W

Wang, M., 68
Wang, R., 18, 69, 97, 135
Wu, M., 106, 124

X

Xie, Q., 17, 18, 27, 28, 44, 45, 59, 61, 149
Xiong, C., 15, 17, 21, 27, 129, 130
Xu, A., 68
Xu, H., 14, 130, 152, 154
Xu, P., 130, 152
Xu, X., 18, 69

Y

Yang, S., 15, 18, 59, 152

Z

Zarrin, S., 15, 64, 65
Zhang, C., 106, 124, 136
Zhang, F., 153
Zhang, L., 18, 59, 61
Zheng, C., 124

Subject Index

A

Accelerated life testing (ALT), 1, 2, 68, 101, 106
Algorithm
 Gauss-Newton, 57
 Newton-Rapshon, 18, 41, 43, 48, 53, 57, 58, 60, 63–65, 76, 97
Asymptotic normality, 43, 80, 83, 93

B

Bayes estimator/estimate (BE), 17, 18, 31, 34, 36, 37, 50, 52–54, 67, 68, 70, 72, 73, 81–83, 92, 94, 101, 133, 141, 142
Bayesian analysis, 31, 68, 101, 129, 155
Bayesian inference, 17, 26, 29, 31, 45, 46, 49, 54, 58, 68, 71, 83, 127, 129, 133, 140, 143

C

Censoring
 Generalized Hybrid Type-I (GHCS-I), 4, 5
 Generalized Hybrid Type-II (GHCS-II), 4, 5
 Hybrid Type-I (HCS-I), 4, 5, 8, 17, 28
 Hybrid Type-II (HCS-II), 4, 5, 9, 17, 28
 Progressive, 56, 106, 148, 149
 Progressive hybrid, 136
 Progressive Type-I (PCS-I), 6, 92, 148
 Progressive Type-II (PCS-II), 7, 9, 17, 28, 90, 93, 149
 Progressively hybrid Type-I (PHCS-I), 92
 Type-I, 3, 4, 8, 18, 56, 59, 80, 83, 84, 87, 92, 94, 97, 101, 106, 114, 130, 137, 145, 146, 150, 152
 Type-II, 3, 4, 8, 17, 18, 20, 27, 29, 45, 56, 66, 79, 80, 86, 88, 89, 106, 114–116, 130, 137, 146, 152

Unified hybrid, 28
Competing risks, 15, 105–107, 110, 112, 114–116, 120, 122, 125, 126, 129, 133, 149, 150, 152, 155
Complementary risks, 15, 106, 125–127
Confidence interval, 15, 23, 26, 27, 37, 43, 44, 52, 58, 61, 65, 67, 71, 88, 89, 95, 120, 122, 127, 129, 136, 140, 142
 Asymptotic, 24, 25, 31, 36, 43, 44, 49, 53, 58, 70, 87–89, 91, 93, 94, 119, 120, 122, 124, 138, 140
 Bootstrap, 25, 26, 31, 36, 43–45, 49, 58, 71, 77, 91–94, 120, 122, 133, 136, 140
 Exact, 17, 18, 21–24, 26, 28, 31, 43, 49, 70, 71, 79, 114, 119, 120, 131–133, 136
Confidence region, 144
Consistent/consistency, 34, 35, 44, 50, 51, 80, 83, 94, 153
Credible interval (CRI), 31, 34–37, 51, 52, 70, 73, 92, 94, 101, 133, 141
 Height posterior density (HPD), 36, 37, 50–53, 68, 142
 Symmetric, 37, 73, 74

D

Distribution
 Birnbaum-Saunders, 15, 19, 55, 65, 66, 68
 Dirichlet, 71, 72, 100, 101
 Exponentiated exponential, 18, 46
 Exponentiated Weibull, 46, 54
 Gamma, 15, 17–19, 22, 37, 46, 49, 51, 55, 56, 58, 59, 79–82, 91, 105, 118, 132, 133, 142
 Generalized exponential, 15, 18, 19, 37, 46, 47, 52–54, 59, 89, 106, 109, 127
 Generalized Weibull, 89, 90
 Generalizes extreme value, 98

Geometric, 15, 18, 19, 68, 69, 71

Log-logistic, 19, 55, 61, 62

Log-normal, 15, 18, 19, 37, 55, 58, 59, 62, 89, 94, 105, 152, 153

One-parameter exponential, 15, 18, 19, 26, 28, 37, 39, 55, 69, 74, 80, 83, 85, 91, 97, 105, 106, 108, 109, 114, 120, 129, 130, 133, 136, 137, 142–145, 147–152, 155

Pareto, 15, 19, 55, 63–65, 89, 92, 93

Two-parameter exponential, 15, 18, 19, 26, 28, 31, 37, 46

Weibull, 12, 13–15, 17–19, 26, 37–39, 45, 46, 48, 49, 54, 55, 58, 59, 79, 80, 83, 95, 97, 105, 106, 108, 109, 122, 124, 125, 129, 133, 151–153, 155

F

Function

Conditional moment generating (CMGF), 21, 30, 116, 117, 142

Cumulative distribution (CDF), 1, 9–14, 17, 19, 26, 28, 35, 37, 39, 46, 47, 53, 55, 56, 58, 59, 61–66, 68, 69, 74, 79, 90, 95, 101, 107, 110, 112–114, 122, 123, 126, 133, 136, 150, 153

Failure rate (FRF), 13, 95, 97

Hazard, 13, 14, 19, 20, 37–39, 46, 47, 55, 56, 59, 62–64, 66, 68, 69, 90, 99–101, 105, 107–109, 112, 113, 121, 125, 153

Moment generating (MGF), 90, 132

Probability density (PDF), 12, 19, 21, 22, 28, 30, 31, 34, 35, 37–40, 46, 47, 49, 50, 55, 56, 59, 61–66, 71, 73, 75, 76, 80–84, 89–95, 97–101, 107–115, 117, 118, 122, 123, 125, 126, 131–134, 136, 137, 140

Probability density(PDF), 91

Probability mass (PMF), 117

Reversed hazard, 125, 126

I

Importance sampling, 18, 34, 36, 50

L

Latent failure lifetime, 106, 110, 112, 122, 126, 127, 152

M

Markov chain Monte Carlo (MCMC), 92, 94, 99, 101, 154

Maximum likelihood estimator/estimate (MLE), 15, 17, 18, 21, 22, 24–29, 31, 36, 37, 40, 43, 44, 48, 49, 52, 53, 56–58, 60, 61, 63–65, 67, 68, 70, 71, 76, 77, 79–81, 83–89, 91–97, 102, 103, 106, 114–116, 118–120, 122–124, 127, 129–136, 138–140, 142, 143, 147, 149–153

Model

Cox's hazard rate, 13

Cumulative exposure (CEM), 10, 11–15, 17–20, 28, 37, 39, 47, 56, 59, 62, 64, 66, 69, 74, 79, 99, 105, 106, 110, 112, 114, 122, 126, 127, 131, 136, 137, 142, 144–146, 148–153

Cumulative risk (CRM), 14, 15, 79, 99, 101, 105, 106, 113, 120, 121

Latent failure time, 105, 106, 107, 125, 133, 150, 152

Proportional hazard rate, 153

Proportional reversed hazard rate, 153

Tampered failure rate (TFRM), 13, 14, 15, 79, 95, 97–99, 105, 106, 112, 123, 133, 152, 153

Tampered random variable (TRVM), 11, 12–15, 79, 80, 83, 92, 94, 99, 105, 150

Weibull proportional hazard model, 14

O

Optimum

A-optimum, 130, 143, 146, 150, 153

C-optimum, 144, 145–147

D-optimum, 130, 144, 146, 148–153

P

Posterior distribution, 34, 50, 67, 72, 73, 81, 82, 91, 92, 94, 99, 101, 140, 141